# Big Data

**Management and Analytics**

# World Scientific Series on Future Computing Paradigms and Applications

**Series Editor: Brij B Gupta** (*Asia University, Taichung, Taiwan*)

---

*Published:*

Vol. 1    *Big Data Management and Analytics*
       by Brij B Gupta and Mamta

World Scientific Series on
Future Computing Paradigms
and Applications – **Vol. 1**

# Big Data
## Management and Analytics

Brij B Gupta
Asia University, Taiwan

Mamta
Punjab Engineering College, Chandigarh, India

**World Scientific**

NEW JERSEY · LONDON · SINGAPORE · BEIJING · SHANGHAI · HONG KONG · TAIPEI · CHENNAI · TOKYO

*Published by*

World Scientific Publishing Co. Pte. Ltd.

5 Toh Tuck Link, Singapore 596224

*USA office:* 27 Warren Street, Suite 401-402, Hackensack, NJ 07601

*UK office:* 57 Shelton Street, Covent Garden, London WC2H 9HE

**Library of Congress Cataloging-in-Publication Data**

Names: Gupta, Brij, 1982–    author. | Mamta, author.
Title: Big data management and analytics / Brij B Gupta, Asia University, Taiwan,
    Mamta, Punjab Engineering College, Chandigarh, India.
Description: New Jersey : World Scientific, [2024] |
    Series: World Scientific series on future computing paradigms and applications ; vol. 1 |
    Includes bibliographical references and index.
Identifiers: LCCN 2023017153 | ISBN 9789811257117 (hardcover) |
    ISBN 9789811257124 (ebook for institutions) | ISBN 9789811257131 (ebook for individuals)
Subjects: LCSH: Big data. | Database management. | Data mining.
Classification: LCC QA76.9.B45 G86 2024 | DDC 005.7--dc23/eng/20230601
LC record available at https://lccn.loc.gov/2023017153

**British Library Cataloguing-in-Publication Data**
A catalogue record for this book is available from the British Library.

For any available supplementary material, please visit
https://www.worldscientific.com/worldscibooks/10.1142/12869#t=suppl

Desk Editors: Sanjay Varadharajan/Amanda Yun

Typeset by Stallion Press
Email: enquiries@stallionpress.com

*Dedicated to my parents and family for their
constant support during the course of this book.*
— **Brij B. Gupta**

*Dedicated to my parents, beloved husband, my son and my mentor
for their motivation throughout the journey of completion of this book.*
— **Mamta**

# Foreword

In the dynamic and ever-changing world of data management and analytics, *Big Data Management and Analytics* stands out as a prominent source of lucidity and a vast reservoir of expertise. In the current epoch of digital transformation, it is of utmost importance to comprehend the fundamental principles and complexities associated with big data. This book serves as an invaluable resource, offering the necessary guidance to attain this objective.

Within the domain of big data, characterized by an overwhelming abundance of information in terms of volume, velocity, and variety, this literary work emerges as a beacon of guidance, catering to a wide-ranging audience. This work is expected to provide significant benefits to a wide range of individuals, including students, data analysts, data scientists, and IT professionals. What is particularly noteworthy about this is that it operates under the assumption that the user possesses no prior understanding or expertise in the field of big data management and analytics. As a result, it is designed to be easily understood and utilized by individuals at all levels of proficiency, including those who are new to the subject matter as well as those who have extensive experience in the field.

The book's structure is characterized by the inclusion of 10 carefully constructed chapters, each designed to guide readers through a comprehensive journey. The initial chapter serves as an introduction to big data, providing readers with a foundational understanding of the subject matter. This introductory chapter serves the purpose of elucidating the concept of big data, aiming to dispel any misconceptions or uncertainties surrounding it. Additionally, it delves into the diverse origins of big data, its distinctive attributes, and the various challenges that arise in relation to its effective management and analysis. Furthermore, this provides valuable insight into the practical implementations of big data in various industries and sectors.

The subsequent chapters of this research work delve into a comprehensive examination of Big Data Management and Modelling, Big Data Processing, Big Data Analytics and Machine Learning, and Big Data Analytics Through Visualization. The aforementioned chapters not only explicate the theoretical concepts but also offer valuable practical perspectives on effectively tackling the challenges that manifest within these specific domains.

The chapter titled "Taming Big Data with Spark 2.0" provides an in-depth exploration of Apache Spark, emphasizing its architectural design and constituent elements. This study aims to provide a comprehensive understanding of the significance of Resilient Distributed Datasets (RDDs) and their utilization in the processing and analysis of large-scale datasets, commonly referred to as Big Data. The acquisition of this knowledge is of utmost significance in contemporary society where the utilization of real-time analytics and processing holds great prominence.

The chapter titled " Managing Big Data in Cloud Storage" provides a thorough examination of data storage systems, specifically focusing on Hadoop Distributed File System (HDFS) and cloud storage. This chapter aims to equip readers with the necessary understanding to effectively navigate the complexities associated with storing data in the cloud.

The book's utility is further expanded through an in-depth exploration of specific industry applications, which serves as a testament to its practical significance in real-world contexts. The chapters titled "Big Data in Healthcare" and "Big Data in Finance" offer valuable insights into the significant impact of big data within these respective industries. The following topics will be explored in this text: personalized medicine, clinical decision support, population health management, fraud detection, risk management, trading strategies, and customer segmentation.

The concluding chapter, titled "Enabling Tools and Technologies for Big Data Analytics," presents a comprehensive collection of essential tools and technologies that are imperative for individuals aiming to leverage the complete capabilities of big data.

The publication accommodates a wide range of readers, rendering it appropriate for individuals with diverse educational and occupational backgrounds. The provided resource employs a systematic methodology to ensure that individuals acquire a comprehensive comprehension of the big data ecosystem, starting from its foundational elements and progressing incrementally. In addition, the book presents various real-world application scenarios that serve to provide readers with the requisite skills and knowledge required to effectively tackle the complexities associated with the management and analysis of big data.

This book will be regarded as an invaluable resource by publishers and industry professionals alike. In the current era characterized by the widespread adoption of data-driven decision-making practices, the content presented in this book serves as a valuable resource for organizations seeking to harness the full potential of their data. By leveraging the insights and methodologies outlined within, organizations can effectively enhance their decision-making processes, leading to improved performance outcomes and ultimately gaining a competitive edge in their respective industries.

In summary, *Big Data Management and Analytics* is a highly valuable resource that merits inclusion in the personal libraries of individuals with a keen interest in the field of data management and analytics. Presented before us is a remarkable occasion to commence a voyage that shall profoundly influence our comprehension of data and unlock avenues to prospects that were previously unattainable. The commencement of the expedition is now underway.

**Prof. Jinsong Wu**

*Chair, IEEE Technical Committee of Big Data*

*Professor, School of Artificial Intelligence, Guilin University of Electronic Technology, 510004, China*

*Department of Electrical Engineering, University of Chile, 8370451, Chile*

# Preface

In today's digital age, data has become an integral part of our lives, both personally and professionally. The amount of data generated every day is massive, and traditional methods of data management and analysis are no longer sufficient to handle this volume of information. This is where big data management and analytics come into play. Big data management refers to the efficient handling, storage, and processing of large and complex datasets, while big data analytics involves the use of advanced analytical techniques to extract valuable insights and patterns from this data. Together, they have the potential to transform businesses and industries, enabling organizations to make data-driven decisions and gain a competitive edge.

This book on *Big Data Management and Analytics* aims to provide a comprehensive and practical guide to the principles, technologies, and tools involved in the effective management and analysis of big data. It covers a range of topics, including data modeling, data storage, data processing, data analysis, and data visualization, including the relevant applications and use cases.

This book is designed for a broad audience, including students, data analysts, data scientists, and IT professionals. It assumes no prior knowledge of big data management and analytics and provides a step-by-step approach to learning the key concepts and techniques involved. By the end of the book, readers will have a thorough understanding of the big data ecosystem and be equipped with the skills and knowledge necessary to effectively manage and analyze big data in real-world scenarios.

This book contains 10 chapters, with each chapter covering different aspects of big data management and analytics. We provide a detailed overview of the topics covered by each chapter in the following:

**Chapter 1: Introduction to Big Data** — This chapter introduces the concept of big data and discusses various sources of big data, its characteristics, and the challenges associated with its management and analysis. It also provides an overview of the popular use cases of big data.

**Chapter 2: Big Data Management and Modeling** — This chapter provides a comprehensive overview of the various techniques and strategies used to manage and model large datasets. Further, this chapter covers the challenges associated with managing and modeling big data.

**Chapter 3: Big Data Processing** — This chapter provides an overview of big data processing, including its definition, characteristics, and challenges. It covers the various requirements for big data processing and discusses in detail the big data processing pipeline. Finally, it discusses the role of Splunk and Datameer in big data processing.

**Chapter 4: Big Data Analytics and Machine Learning** — This chapter covers the process of collecting and preprocessing data for big data analytics. It further covers data transformation techniques and data quality issues that need to be addressed before analysis. It also provides an overview of the role of machine learning in big data analytics and the challenges of implementing machine learning on big data, including scalability and data preprocessing.

**Chapter 5: Big Data Analytics Through Visualization** — This chapter discusses the role of visualization in big data analytics. It covers the various types of visualizations that can be used to explore and analyze big data with special attention to graph analytics.

**Chapter 6: Taming Big Data with Spark 2.0** — This chapter provides an overview of Apache Spark. It discusses the key features and benefits of Spark 2.0, such as the architecture of Spark 2.0, including its components, such as Spark Core, Spark SQL, Spark Streaming, and MLlib. It discusses the role of resilient distributed datasets (RDDs) in Spark 2.0 and how they can be used to process and analyze big data.

**Chapter 7: Managing Big Data in Cloud Storage** — This chapter focuses on the storage and retrieval of big data. It covers the different types of data storage systems, such as Hadoop distributed file system (HDFS), Hue, and cloud storage.

**Chapter 8: Big Data in Healthcare** — This chapter provides a comprehensive introduction to the role of big data in healthcare. It further discusses the architectural framework of big data in healthcare. Finally, it covers some common applications of big data in healthcare, such as personalized medicine, clinical decision support, and population health management.

**Chapter 9: Big Data in Finance** — This chapter discusses the role of big data in finance and how big data is transforming the financial sector. It covers the various sources of financial data, such as transaction data, market data, social media data, and news feeds, and throws light on the challenges associated with processing and analyzing such data. The chapter also details some common applications of big data in finance, such as fraud detection, risk management, trading strategies, and customer segmentation.

**Chapter 10: Enabling Tools and Technologies for Big Data Analytics** — This chapter provides a comprehensive introduction to various tools and technologies that enable big data analytics. It covers the tools and technologies for big data management, modeling, integration, processing, analytics and visualization.

# About the Authors

**Brij B. Gupta** is the Director of the International Center for AI and Cyber Security Research and Innovations and a Distinguished Professor at the Department of Computer Science and Information Engineering (CSIE), Asia University, Taiwan. In more than 17 years of his professional experience, he has published over 500 papers in journals/conferences, including 35 books and 12 Patents with over 25,000 citations. He has received numerous national and international awards, including the Canadian Commonwealth Scholarship (2009), the Faculty Research Fellowship Award (2017), the Visvesvaraya Young Faculty Research Fellowship Award from the Indian Ministry of Electronics and Information Technology, Government of India, the IEEE GCCE outstanding and Women in Engineering (WIE) paper awards, and National Institute of Technology Kurukshetra, India's Best Faculty Award (2018 and 2019), respectively. Prof. Gupta was selected for Clarivate's Web of Science Highly Cited Researchers in Computer Science (top 0.1% researchers in the world) consecutively in 2022 and 2023. He was also listed in Stanford University's ranking of the world's top 2% of scientists in 2020, 2021, 2022, and 2023.

Dr Gupta is also a visiting/adjunct professor with several universities worldwide and an IEEE Senior Member (2017). He was selected as the 2021 Distinguished Lecturer in IEEE Consumer Technology Society (CTSoc). Dr Gupta is also serving as a Member-at-Large of IEEE

Consumer Technology Society's Board of Governors (2022–2024). He is also the Editor-in-Chief of the *International Journal on Semantic Web and Information Systems, International Journal of Software Science and Computational Intelligence, Sustainable Technology and Entrepreneurship,* and *International Journal of Cloud Applications and Computing,* the Lead Editor of a Book Series with CRC and IET Press, and the Associate/ Guest Editor of various journals and transactions. He also served as a member of the technical program committee of more than 150 international conferences. His research interests include information security, cyber physical systems, cloud computing, blockchain technologies, intrusion detection, AI, social media, and networking.

**Mamta** received her BTech (with honors) in computer engineering in 2011 from the University Institute of Engineering and Technology, Kurukshetra University, India, and her MTech in computer engineering (with distinction) in 2014 from the National Institute of Technology, Kurukshetra, India. She received her PhD in computer engineering from the National Institute of Technology, Kurukshetra, India, in 2020. At present, Dr. Mamta is an assistant professor in the Department of Computer Science and Engineering at Punjab Engineering College, Chandigarh, India. Her research interests include applied cryptography, searchable encryption, information security, big data security, and cloud computing. She has published several research articles with various reputed publishers, including IEEE, Springer, and Wiley.

# Acknowledgments

First and foremost, we would like to express our deepest and most sincere gratitude to the Almighty for granting us the strength and guidance to complete this book. We would like to thank our families for their unwavering support and encouragement. We are also grateful to our friends and colleagues who have provided invaluable feedback and helped us refine our ideas.

Further, our deepest appreciation goes to the team at World Scientific Publishing, who believed in this project and worked tirelessly to bring it to fruition.

Finally, we would like to acknowledge the countless individuals who have shared their knowledge and expertise with us through interviews, research, and personal interactions. Their insights have been invaluable in shaping this book and we are deeply grateful for their contributions.

We thank all for the support and encouragement, without which this book would not have been possible.

# Contents

# List of Figures

# List of Tables

# Chapter 1

# Introduction to Big Data

Digitalization in the modern world leads to the generation of a large amount of data in almost every domain. Due to the sheer volume of data that needs to be managed, big data management and analysis have become essential components of any system. The large amount of data that is being produced as a result of the proliferation of connected devices is growing at an exponential rate, which makes it extremely difficult to store and process all of this data. The challenges associated with storing, transmitting, and understanding this data are so enormous that they have given rise to their very own field of study as well as an industry: *Big Data*. The management and analysis of large amounts of data have become increasingly popular in virtually all sectors, including the healthcare and financial sectors. Now, the big question is: What actually triggered the big data era? First is the continuously growing torrent of data, and the second is the on-demand cloud computing services. The data are generated continuously at a very fast pace. According to a report by McKinsey Global Institute (MGI), worldwide data volumes are doubling every two years due to social websites, sensors, smartphones, and other sources [1]. The exponential increase of data is fueled by the rising volume and detail of data acquired by businesses, the advent of multimedia, social media, and the Internet of Things. This exploding data can add value to businesses, and there have been instances where with proper analytics, big data reveal useful insights and become a critical factor in almost every industry. MGI studied the impact of big data in several domains like healthcare, retail, manufacturing and the public sector and noted the instances where big data is contributing value to each of these domains [1]. The biggest source of data is mobile devices, which every day and all the time generates data. According to a recent report by Statista, by the end of 2021, the number of mobile devices in use globally will be over 15 billion, and by 2025, the number of mobile devices is predicted to reach 18.22 billion, a 4.2 billion increase above 2020 levels [2], and so will be the data. This serious growth in data demonstrates the immediate need for information on how to handle and manage such data. Apart from this continuously growing data, there is another factor which contributed to the launch of the big data era, and that is cloud computing. Cloud computing is often called on-demand computing. It allows us to perform computing anytime and anywhere, which, when combined with this continuously growing torrent of data, opens new gates for dynamic and scalable data analysis. This novel and dynamic data analysis have the ability to reveal insights about the world which were difficult to imagine before. Hence, cloud computing and the growing torrent of data together launched the new era of Big Data.

## 1.1  Data: The New Oil and the New Soil

The phrase "data is the new oil" was originated by Clive Humby, a British mathematician and data science entrepreneur, in 2006, and has subsequently been repeated by numerous others. Since that day, this term has been the subject of several articles, as well as appearing increasingly frequently in Google searches, as seen in Figure 1.1.

This metaphor shows the importance of data. As we know, oil is one of the world's most valuable commodities, which contributes a major part to the global economy. Like oil, data is also a valuable resource, but only if we are able to extract value from it. Like oil, data needs to be refined so that we can run analytics on it to get useful insights about and around us. The phrase "data is the new oil" is further extended by Peter Sondergaard in 2011. According to Peter Sondergaard, if data is the new oil, then

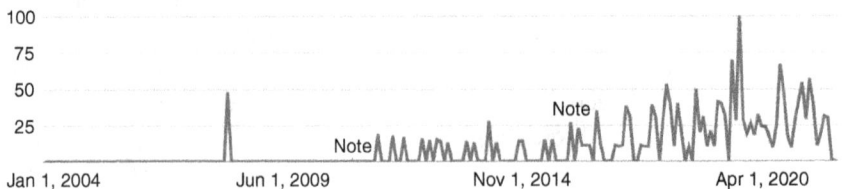

**Figure 1.1.**   Google Searches for the Phrase "Data is the New Oil" [3]

analytics is the combustion engine, which actually converts the data into a form that can be consumed directly by the 21st-century business world.

In the 21st century, data is equivalent to what oil was in the 18th century, the driving force of the economy. But unlike oil which we consume, data is something that we generate. Data, like soil, is a fertile medium that can be improved and reused over time. It is virtually infinite, and it will not exhaust as we use it. In reality, data-driven understanding has allowed us to create unlimited, inexhaustible, and healthy renewable energy-like sources. Hence, soil data is an unlimited and essential element for survival in the present business world. Further, there is another aspect in which we see the importance of soil. It corresponds to the land that a particular individual or a country owns, and that is a symbol of power and prestige. Everywhere around us, we see people and even countries fighting for it. A similar kind of importance can also be given to data. In the present digital economy, data is equivalent to oil and soil. The one who is able to manage, handle, analyze and get value or insights out of the data can enter into the race to become the world's leader.

## 1.2 What is Big Data and What are its Sources

Big data generally refers to large volumes of data. It is the data that is too big, too fast, or too hard for existing tools to process [4]. The terms "too big", "too fast" and "too hard" refer to the three popular V's of big data, i.e., volume, velocity and variety, which we will discuss in the following section on characteristics of big data. Now, the question is: Where does this big data come from? There are three major sources of big data: machines, humans, and organizations, as shown in Figure 1.2 [5].

Machine data refers to the data generated by industrial sensors, environmental sensors, sensors used in smart vehicles, personal health trackers, etc. One example of machine-generated data is the Large Hadron Collider, the largest and most powerful particle collider in the world, which generates 40 terabytes of data every second. The human-generated data refers to social media data, tweets, photos, status updates, and other media. The organizational data refer to transaction history, logs, customer information, etc. According to a report by Internet Data Centre (IDC), the worldwide data contributed by different sources will rise to 175 zettabytes by 2025, where 1 zettabyte equals to $1000^7$ bytes [6]. So, in the zettabyte era, we can imagine the volume of the data that need to be processed to get useful insights. Further, in most business use cases, any single source

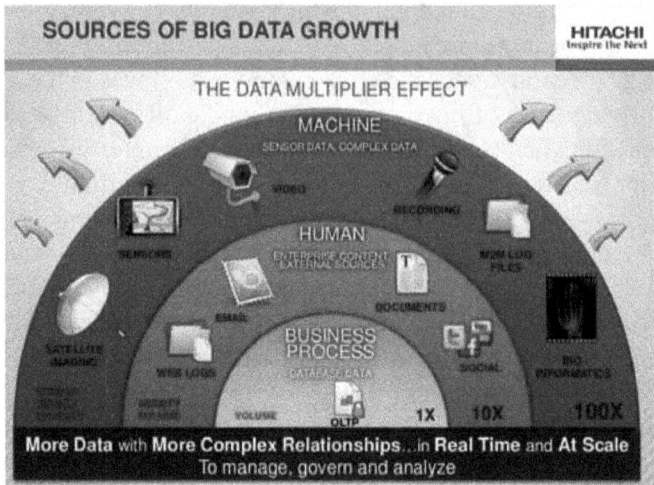

**Figure 1.2.**   Big Data Sources [5]

of data is never sufficient. The real value can only be extracted by combining different streams of these big data sources and analyzing them to generate valuable insights.

## 1.2.1 Big Data Generated by Machines

Machine-generated data is everywhere around us, and it is the largest among all the big data sources. For example, a Boeing jet aircraft produces around half a terabyte of data every time it takes off. There are tremendous sensors implanted in its body which continuously generate data and send its current status to the teams on the ground and on the plane. Machines are equipped with smart sensing capabilities, which result in data generation. For example, a cellphone, which we now call a smartphone, is equipped with sensors like motion sensors, environmental sensors, proximity sensors, and a lot more. We use the term "smart" for the cellphone because it enables us to track and connect to other things. In general, we call a device a smart device based on three main properties: the ability to connect to other devices or the network, execute services and autonomously collect data, and have some knowledge of the environment. Smart devices are widely available, and they are often interconnected, which led to something we call today the Internet of Things (IoT).

These connected smart devices in IoT networks are continuously generating data, and these data constitute what we call machine-generated data. Machines generate data 24/7 through the in-built sensors, both at a personal and an industrial scale, and thus contribute a major proportion to the big data among all sources. The biggest advantage of machine-generated data is that it enables real-time data analysis, which results in real-time monitoring, notification and action.

## 1.2.2 Big Data Generated by Humans

People produce enormous amounts of data on a daily basis as a result of their use of social media websites, such as Facebook, Instagram, LinkedIn, and Twitter, and online video-sharing websites, such as YouTube. These websites allow users to upload, share, and view photographs and videos. In addition to these sources, a heavy amount of data is generated by performing internet searches, writing blogs, making comments, writing text messages and personal emails. A glimpse of these data sources is shown in Figure 1.3. One of the biggest drawbacks of data generated by humans is that the majority of the data is highly unstructured and is non-adhering to well-defined data models.

Table 1.1 shows the daily data volume figures from some of the leading internet platforms; these statistics are interesting since they fall within the Petabyte range for everyday tasks.

This vast volume of primarily unstructured data produced by people poses numerous challenges. Unstructured data do not conform to the

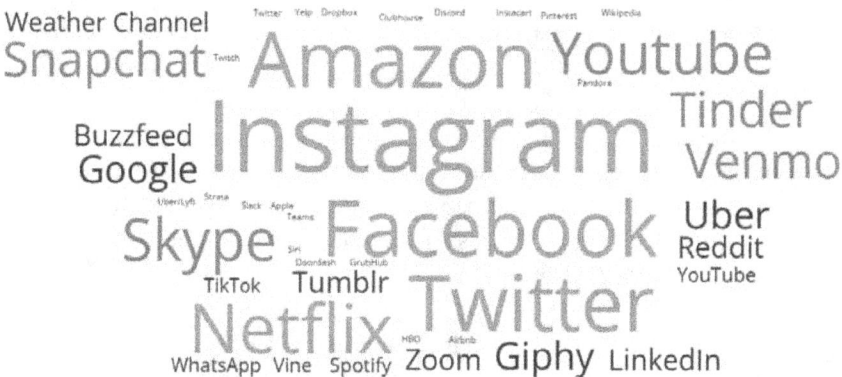

**Figure 1.3.** Human-Generated Big Data Sources [7]

**Table 1.1.**  Daily Data Volumes for Popular Online Platforms [8]

| Company | Amount of Data Generated Each Day |
|---|---|
| Google | 100 Petabyte (PB) (1 PB = 1000 TB) |
| Facebook | 30 + PB |
| Twitter | 100 Terabyte (TB) (= 0.1 PB) |
| Spotify | 64 TB |
| eBay | 100 PB |

pre-defined data models, i.e., these constitute data that we cannot store in traditional database systems. We cannot use the relational database management system (RDBMS) to manage and extract this kind of data. As an example of structured data, consider a sales receipt that we get when we buy something: the invoice has a proper structure, and there is a date section, name section, list of items purchased, and the total amount to be paid. While on the other hand, unstructured data relates to data that humans generate in the form of text that we write with our hands. These do not follow any well-defined structure, and out of the entire data in the world, 80–90% relate to unstructured data, and this number is growing at a very fast pace, and technically, this is what we call the velocity of the big data. The following information graphics shown in Figure 1.4 can give us the idea of the data that we contribute each day and every single minute on the Internet.

In addition to text, people also create various forms of unstructured data, such as video, audio, photographs, web searches, and emails. It is difficult to manage and analyze unstructured data since it might be in any of the wide variety of formats designed primarily for human consumption, such as HTML, pictures, PowerPoints, PDFs, and so on. It may take a considerable investment in time and money to collect, store, cleanse, retrieve, and evaluate unstructured data before we can start enjoying its benefits. Further, the tools and people needed to carry out the extraction process from these huge volumes of unstructured data are quite difficult to find. However, organizations are still using this kind of data to find out useful insights and trends.

Now, the question is how data generated by people is being used. What are the emerging technologies that can tackle these challenges? The Hadoop open-source big data framework is the backbone of most of the

**Figure 1.4.**   Data Generated in Each Minute over the Internet [9]

unstructured big data management tools available today. Hadoop handles large datasets in a distributed manner and thus tackles the first challenge, namely the volume of the unstructured data. But in addition to this, there is a need for real-time data processing to handle the data from sources, such as Twitter and Facebook. The other two open-source frameworks that can deal with such rapidly produced real-time data are Spark and Storm. Every database or data storage system is compatible with both of these tools. Regrettably, in today's dynamic big data world, we cannot utilize standard RDBMS to store this unstructured data because it is too static. Most modern enterprises, however, have discovered a workaround for this issue by adopting a hybrid strategy whereby smaller, structured datasets are kept in relational databases, and larger, unstructured datasets are maintained in cloud-based NoSQL databases. NoSQL Data technologies, in contrast to the traditional relational database-centric warehouses, are based on non-relational notions and provide data storage solutions frequently in the form of cloud computing. For instance, a graph database is an optimal choice when the data will be used in an analysis to reveal connections between different types of data. The graph database Neo4j serves as an example. There are certain databases designed for the purpose of storing and retrieving data using keys and values, such as in a search engine. Cassandra is an example of a key-value database. Apart from these two, there are several other types of NoSQL systems that can address individual challenges.

The next question is how to use these emerging technologies to generate valuable insights from people-generated big data. As we know, to generate value out of data, it needs to be passed through several steps, such as storage, retrieval, cleaning, and analysis. It is possible to find a solution to this issue by running each stage as a separate layer, adapting the currently accessible tools to the nature of the issue at hand, and scaling analytical methods to large amounts of data. Organizations listen to the real voice of customers using big data, and one such example is sentiment analysis. In this type of analysis, companies examine social media and other data to determine whether people have a good or negative perception of their company. Personal data processing is being used by businesses to better understand their consumers' genuine preferences. Twitter analyzes around 12 terabytes of data every day to find out sentiments around some product or topic. To compare, it would take two nonstop years to listen to one terabyte of music. There are other such examples where companies, such as Amazon and Netflix, are using people-generated data to determine the preferences of their customers, to provide them best offers, and to recommend products and services based on their past likings, thereby resulting in happy customers and increased profits.

### 1.2.3 Big Data Generated by Organizations

The big data generated by organizations is structured, but it comes in silos. The generation of data by an organization depends on the context, and it varies from organization to organization. Since every organization has its own unique set of operational procedures and marketing strategies, it results in a diverse range of data production platforms. A bank, for example, receives significantly different types and sources of data than a hardware equipment maker. Some prevalent sources of organizational big data are the transactions performed by users, e-commerce, stock records, medical records, etc. Organization store their data for present or future use and perform analysis of the past. For example, an organization collects sales data and uses pattern recognition to find connected products, antici- pate demand for products that are expected to increase in sales, and catch fraudulent activities. Organizations typically store every event of their interest and run analytics to maintain their inventories as per the predicted demands. Traditionally, most organizations manage and process highly structured data for their operational and business intelligence systems, and as a result, RDBMS was highly adopted by most of them. However, it is still challenging to integrate this structured data as there is a continuum of technologies which model, collect and process data coming from different software and hardware components within an organization. Such issues in the past resulted in information being housed in silos, even within an organization, and if such silos are left untouched, the organizations are at risk of being outdated and unsynchronized. Further, most organizations have typically collected data at the departmental level, with no infrastruc- ture or policies in place to exchange and combine information, since there is not a single system within the organization that can access all of the data. The adverse effects of maintaining such a rigid structure are becom- ing increasingly apparent to businesses, who are therefore altering their policies and architecture in order to make room for the integrated process- ing of all data for the greater good of the whole enterprise. In this domain, cloud-based solutions are considered possibilities that are both versatile and cost-effective.

Now, the question is how organizations are benefiting from big data. Organizations can have real benefits from the organization-generated data if they combine it with other types of data. Walmart is an example of a company in the retail sector that makes extensive use of big data. Walmart is a large organization with 250 million customers and 10,000 stores, and they collect around 2.5 PB data per hour. They monitor a wide range of information pertaining to sales, customers, and products, including tweets, in-store transactions, clicks on the company's website, and more. They make use of this data to detect patterns, such as which products are frequently purchased together and which new products to offer in their stores, to estimate demand in a particular area, and to tailor the recom- mendations they provide. Overall, Walmart has retained its position as a leading retailer by embracing big data and analytics. Almost all the leading companies are now utilizing big data, and studies forecast that total spending will drastically increase in the future. According to a study conducted by Bane and Company, early adopters of big data analytics have a huge advantage over the rest of the corporate sector, which is shown in Figure 1.5 [10].

As shown in Figure 1.5, companies that utilize analytics have a five times greater likelihood of making decisions much faster than their peers, a three times greater likelihood of executing decisions as intended, and a two times greater likelihood of using data very frequently when making decisions. Additionally, these companies are twice as likely to be in the

**Figure 1.5.**   Early Adoption Benefits of Big Data [10]

top quartile of financial performance inside their own industry. Integrating big data approaches into a company's culture and breaking down silos have proven to be extremely beneficial. Operational efficiency, higher sales, increased revenues, and improved customer satisfaction include just a few of the primary benefits to organizations.

## 1.3 Characteristics of Big Data

Big data is a blanket term for any collection of data that is huge and complicated enough to exceed the processing capabilities of traditional data management systems and procedures. Big data is commonly characterized by a number of V's, as shown in Figure 1.6.

The first three are Volume, Velocity and Variety, which are coined by Doug Laney [11]. The term "volume" alludes to the enormous quantity of data that is produced every instant, period, and day in this era of digital technology. The pace at which data are generated as well as the rate at which the data move from one point to the next are both referred to as the velocity of data. Variety refers to the several configurations/forms that the

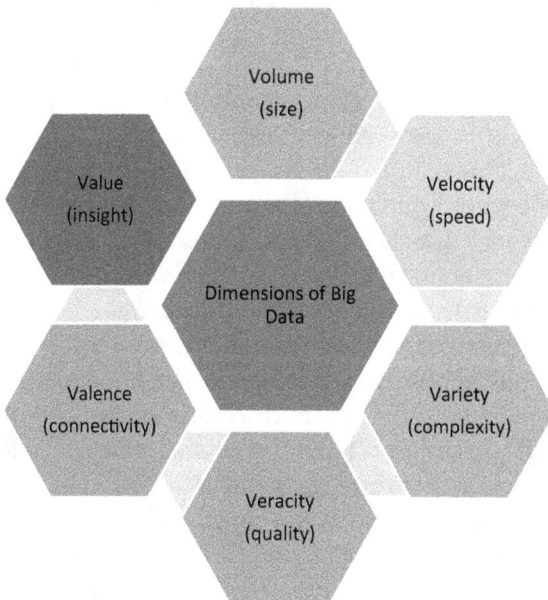

**Figure 1.6.**   The Six V's of Big Data

data are capable of taking. Data can be represented in an extremely wide variety of formats, some examples of which include text, photographs, audio, and geographical data. Big data can be described by its volume, velocity, and variety, which are the three primary dimensions that define it. More and more Vs have been brought to the field of big data as we learn of new challenges and methods for defining "big data". Two of these additional V's are referred to as veracity [12–14] and valence [14]. When talking about data, veracity relates to things like biases, noise, and irregularities. Veracity, to use the term in its most relevant use, refers to the immeasurable degrees to which the accuracy and reliability of data might vary. Valence is the process of connecting vast amounts of data in the form of graphs. The sixth V is the Value [14], which describes how big data benefits you and your organization.

### 1.3.1 Volume

The amount and rate of expansion of big data are also factors that go into the volume dimension. This volume can come from massive datasets or the collection of many little data pieces and events over time. The size and scope of big data storage can be enormous. According to a report by IDC, there will be a 61% increase in the amount of data stored digitally and this will become around 175 zettabytes [6], where a zettabyte is 1 trillion gigabytes, i.e., $10^{21}$. One can only fathom the time span, money, and effort that will be required to store such a large amount of data, and understanding the criticality of these is a little difficult. After that, the following era will be of yottabytes ($10^{24}$) and brontobytes ($10^{27}$) which is pretty impossible for us to comprehend at this moment. This is called data growing at an astronomical scale. The huge volumes of big data come with a variety of challenges. Certainly, storage is the most apparent one. The larger the amount of data to be stored, the greater the amount of storage space that will be necessary to store the data effectively. We will need to have the capability to quickly retrieve this enormous volume of data and then promptly transmit it to the processing units. This leads to additional issues, such as networking, bandwidth, the cost of data storage, whether to store data locally or in the cloud, and so on. The processing of such staggering quantities of data poses further complications. In terms of memory, processor, and input/output requirements, most present analytical approaches will not scale to such large amounts of data.

## 1.3.2 Velocity

The rate at which large amounts of data are produced, collected, and digested is referred to as velocity. To enable real-time action, processing data in real time should match the rate at which they are created. This feature makes it possible to personalize the advertising that appears on the websites you visit depending on the content that you have previously searched for, watched, or purchased on these websites. When a company is unable to take advantage of data as it is created or as it is analyzed, it frequently misses out on opportunities. It is possible that the rate at which big data is generated and the ability to evaluate it as it is produced could have a substantial impact on the standard of living of humans. Sensors and other types of intelligent technology that monitor the human body can detect abnormalities in real time and take corrective action immediately, hence these have the potential to save lives. This type of processing is known as real-time processing, and it is very different from its more distant relative: batch processing. During batch processing, significant amounts of data are loaded into massive computer systems all at once and then processed in stages over a period of several days.

Decisions based on outdated data, even if only a few days old, can be disastrous for certain organizations. Organizations that base their decisions on the most recent facts are more likely to succeed. So, in order to acquire the capability of making decisions in real time, it is essential to match the pace of processing with the speed at which information is generated. In addition, the sensor-driven sociocultural environment of the modern day calls for decisions to be made more quickly. As a consequence of this, we are unable to wait around for all of the data to be generated before feeding it into a device. There are many different applications where new data are constantly being generated and need to be combined with existing data in order to make decisions. Some examples of these applications include the planning of disaster responses, decisions regarding real-time trading strategies, and marketing forecasts. As more data are collected, the conclusions drawn from the analysis should evolve to reflect the shifting nature of the data's inputs. Processing in batches the information that was acquired in the past could lead to decisions that paint an inaccurate image of the situation. As a consequence of this, the applications require what is known as streaming analysis, which is a state of the current environment that is updated in real time. We are fortunate to be able to collect the most current information at a much quicker rate and in real time than we were in the past due to the development of low-cost sensors, cellular phones, and social networks. Streaming data provides real-time updates on what's going on. Streaming data has velocity, which means it is created at different speeds, and real-time analysis of such data provides agility and flexibility to optimize the advantages that an organization wants.

### 1.3.3 Variety

It is a form of scalability, but here the term "scale" does not refer to the size of the data. Instead, it refers to diversity. To solve real-life problems, a variety of data need to be collected, stored, and processed. The four primary dimensions of data heterogeneity are structural variety, media variety, semantic variety, and availability variation. Structural variety refers to the difference in the representation of data. For example, NASA's satellite image of wildfires appears quite different compared to the tweets put out by those watching the fire spread. The term "medium variety" relates to the medium through which the data are transmitted. The audio recording of a speech and the written transcript of the same speech, for instance, both contain the same information; but, they are presented in quite different formats. There can also be some data objects that may contain multiple media, like news video, which is a combination of image sequence, audio, and captioned text. Semantic variety relates to how to interpret and perform operations on data. For example, when we measure amounts, we frequently utilize different units. We sometimes also utilize qualitative vs. quantitative measurements. Availability variation can take many forms. For example, data can be available in real time, such as sensor data, or saved, such as health records. Similarly, data can be accessed on a continuous basis, such as data from a traffic camera, or on a periodic basis, such as just when the satellite passes over the region of interest.

Further, a single data object or a group of related data objects, may not be uniform on their own, there could be variety within type too. For example, Emails are a hybrid entity. The email header, which contains fields like Sender, Receiver CC/BCC, Date, and Subject, constitutes a structured component and can be seen in the form of a table. The body of the email contains text, and the attachment may contain some images, files or other multimedia objects.

### 1.3.4 Veracity

The term "veracity" relates to the accuracy of big data. Depending upon the lifetime of the data, sometimes it is also referred to as volatility or validity of data. Veracity is critically necessary for the successful application of big data. owing to the fact that big data might be messy and untrustworthy. It is possible for it to be loaded with biases, oddities, and inaccuracies, making it worthless, and big data analysis outcomes are only as good as the data being analyzed. Big data offers numerous opportunities for making data-driven decisions, but data evidence is only useful if the data is of satisfactory quality. For example, fake reviews about a product can lead to wrong decisions about maintaining stocks. In the context of big data, the term "quality" can be defined in terms of the correctness of the data, the integrity or validity of the data source, and the process by which the data were produced. Furthermore, the relevance of the data in relation to the programs that analyze them is a significant component, which makes context a part of the quality. The ever-increasing volume of big data necessitates quick ways to incorporate it into analytical solutions. However, this makes it difficult to maintain track of data quality. One should be aware of what information was gathered, where it originated from, and how it was examined before being used.

### 1.3.5 Valence

The connectivity of the data is what is meant by the term "valence", and the higher the valence of the data, the more connected it is. The field of chemistry is where the term "valence" was first derived. In chemistry, there are two types of electrons: the core electrons and valence electrons of an atom. Valence electrons are the ones that are responsible for atom bonding. They are found in the outermost shell of an atom, with the greatest amount of energy. The result of the increased valence is a greater sense of bonding, often known as the connection. The same idea is carried over to the definition of valence when it is used in the context of big data. When two data elements are related to one another in some way, then those two data elements are connected to one another. Most of the time, we find that the data elements are linked to one another. For instance, a city is connected to the nation to which it belongs; an employee is connected to the organization in which he or she works; two users of

LinkedIn are connected because they are friends; and so on. Also, there is the possibility that the data are connected in an indirect manner. For instance, two different teachers are linked together because both of them work in the same department. Valence is a measurement that compares the number of data items within a collection that is actually connected to the total number of data items within the collection capable of having connections. The fact that the data connectedness improves with the passage of time is the aspect of valence that is the most important. A data collection with a high valence is denser. As a result, many standard analytic techniques are ineffective. To account for the rising density, more complicated analytical procedures must be used. Further, the dynamic nature of the data makes the situation even more challenging. It is necessary to analyze, characterize, and make projections on how the valence of a connected collection of data will develop over the course of time and volume.

### 1.3.6 Value

We have looked at the five different aspects of the big data problem that pose a challenge: volume, velocity, variety, veracity and valence. But, the most important component of the problem posed by big data is to convert each of these five dimensions into a value that is actually of some use. The goal of processing all of this enormous data is to bring value to the problem at hand.

## 1.4  Importance of Big Data: Popular Use Cases

Big data has applications in almost every field, whether it is finance, business, banking, healthcare, or any other field [15,16]. Here, we discuss one of the popular use cases in healthcare. The use of big data has the potential to save a significant number of lives. One use of big data in the healthcare industry is precision medicine. It is a relatively new branch of medicine that places emphasis on the patient as an individual. In this branch, they analyze the information related to the genetics of an individual. Further, they keep track of surroundings and everyday activities in order to detect or anticipate a health problem early, thereby assisting in the prevention of sickness and offering the proper treatment at the right dose specifically tailored to that individual in the event of illness. To achieve these capabilities, precision medicine makes use of machine-generated data,

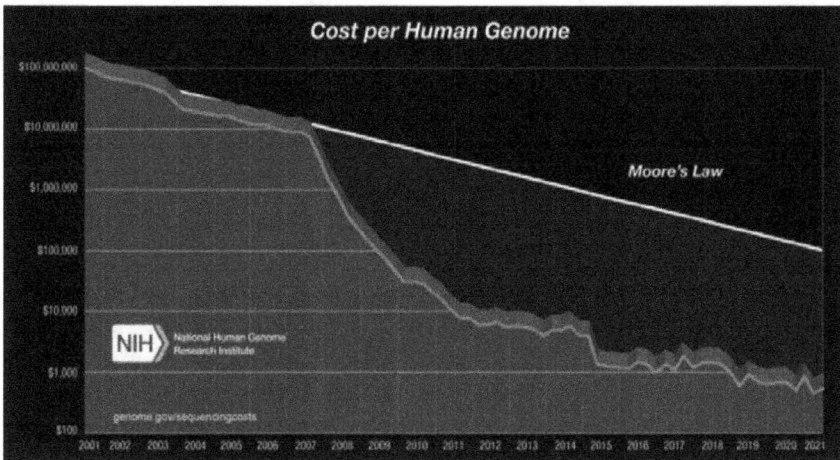

**Figure 1.7.**   Cost per Human Genome [17]

people-generated data and organization-generated data. A crucial component of precision medicine is the utilization of an individual's unique genetic profile in the process of diagnosing and treating that person. As can be seen in Figure 1.7 [17], there has been a significant drop in the amount of money required to analyze the human genome, which holds the key to improving people's overall health.

In addition, there has been a rise in the availability of low-cost, large-scale data storage, and there has also been an increase in the digitalization of records that were previously kept on paper. These are also elements that assist precision medicine. For individual healthcare practices, one additional capability is required, which is the integration of the data from multiple data sources (sensor, people and organization data).

*Machine-generated data*: Digital equipment in most hospitals have been producing data for years, but these data were never ever stored or shared and were intended just for the use of healthcare professionals at that instant and then discarded afterwards. But now these data are stored and analyzed, and further, a new kind of data is becoming quite popular in day-to-day life, and this is the data collected by fitness devices. These fitness devices come in a variety of forms, such as wristbands, watches, shoes, and vests, and they are continually monitoring a variety of indicators by interfacing directly with the smartphone, such as heart rate, blood sugar level, and so on. One can potentially enhance the lifestyle and maintain good health if one keeps a record of the activities one engages in on a daily basis. On the other hand, the fact that the data they collect is about what occurs in regular life as opposed to only what occurs when one goes to the doctor, thus making it potentially very useful information for medical purposes. These devices generate many terabytes of data, which, when combined with information from other sources, such as a digital medical record or a genetic profile, has the potential to make precision medicine a reality.

*Organization-generated data*: The collection of vital scientific data and information for mankind and other transgenic animals at various stages of development has resulted in the establishment of a large number of public databases. These databases include both scientific and mathematical data, which is essential for observations of chronic illnesses, such as cancer, Alzheimer's and Parkinson's disease. In addition, a number of information bases, such as gene ontology [18] and the Unified Medical Language System [19], have been built in order to put biological data in such a structure that can be processed by machines. It is possible that the data on medical care gathered by government agencies could serve as another large source of information.

*People-generated data*: Mobile health applications are becoming increasingly popular. These applications can monitor several vitals, such as cardiac rate, systolic and diastolic pressure, and oxygen levels. Obviously, this is the data generated by sensors [20]. However, there exist several types of mobile applications, which interact with users and take inputs from them [21]. In general, when a patient visits their doctor, they may be asked if they have had any adverse effects from their drugs. In this type of scenario, the accuracy is very less because they might have experienced something several days ago, and they might not remember clearly the exact experience and the reaction. However, nowadays, with the help of these applications, people are self-reporting their responses and experiences. This type of data was never collected before and can be utilized to investigate the effect of the medicine on responses to certain medical conditions at a much more precise and individualized level.

## 1.5 Chapter Summary

In conclusion, big data refers to the vast and complex amount of data that is generated every day by various sources, such as social media, sensors, transactions, and more. The sheer volume, velocity, and variety of this data require specialized tools and technologies to store, process, and analyze it effectively. Further, big data is generated from various sources, including but not limited to social media, machine-generated data, human-generated data, and transactional data. These sources are diverse, and the insights derived from analyzing this data can provide valuable information to organizations to make data-driven decisions and improve their operations. Understanding these sources and how to manage them effectively is critical for organizations looking to leverage big data.

# References

1. Manyika, J., Chui, M., Brown, B., Bughin, J., Dobbs, R., Roxburgh, C., & Hung Byers, A. "Big Data: The Next Frontier for Innovation, Competition, and Productivity", https://www.mckinsey.com/business-functions/mckinsey-digital/our-insights/big-data-the-next-frontier-for-innovation.
2. O'Dea, S. "Number of Mobile Devices Worldwide 2020–2025", https://www.statista.com/statistics/245501/multiple-mobile-device-ownership-worldwide/.
3. Google Trends, https://trends.google.com/trends/.
4. Madden, S. (2012, May–June). From Databases to Big Data. *IEEE Internet Computing*, 16(3), 4–6. doi: 10.1109/MIC.2012.50.
5. Feblowitz, J. I. L. L., Rice, L., Beals, B., & Andersson, B. J. O. R. N. (2013). Big Data in oil and gas: How to tap its full potential. *Hitachi Data Systems Corporation*.
6. Patrizio, A. "IDC: Expect 175 Zettabytes of Data Worldwide by 2025", https://www.networkworld.com/article/3325397/idc-expect-175-zettabytes-of-data-worldwide-by-2025.html.
7. Schein, B. Vice President, Data Curiosity, "Data Never Sleeps", https://www.domo.com/blog/9-years-of-data-never-sleeps/.
8. Neto, J. A. R. (2020). "Measuring the Data Size", https://medium.com/xnewdata/measuring-the-data-size-345320a4ee02.
9. Desjardins, J. "What Happens in an Internet Minute in 2016?" https://www.visualcapitalist.com/what-happens-internet-minute-2016/.
10. https://www.bain.com/insights/who-how-why-big-data-infographic/.
11. Laney, D. (2001). 3D Data management: Controlling data volume, velocity and variety. Meta Group. *Lakshen, Guma Abdulkhader*, 1–4.
12. Normandeau, K. (2013). "Beyond Volume, Variety and Velocity is the Issue of Big Data Veracity", http://insidebigdata.com/2013/09/12/beyond-volume-variety-velocity-issue-big-data-veracity/.
13. Borne, K. (2014). "Top 10 Big Data Challenges — A Serious Look at 10 Big Data Vs", https://www.datasciencecentral.com/profiles/blogs/top-10-list-the-v-s-of-big-data.

14. Big Data: The 6 Vs You Need to Look at for Important Insights, https://www.motivaction.nl/en/news/blog/big-data-the-6-vs-you-need-to-look-at-for-important-insights.
15. Wang, Y., Liu, Q., Hou, H. D., Rho, S., Gupta, B., Mu, Y. X., & Shen, W. Z. (2018). Big data driven outlier detection for soybean straw near infrared spectroscopy. *Journal of Computational Science*, 26, 178–189.
16. Alsmirat, M. A., Jararweh, Y., Al-Ayyoub, M., Shehab, M. A., & Gupta, B. B. (2017). Accelerating compute intensive medical imaging segmentation algorithms using hybrid CPU-GPU implementations. *Multimedia Tools and Applications*, 76, 3537–3555.
17. "DNA Sequencing Costs: Data", https://www.genome.gov/about-genomics/fact-sheets/DNA-Sequencing-Costs-Data.
18. http://geneontology.org/.
19. https://www.nlm.nih.gov/research/umls/index.html.
20. Plageras, A. P., Psannis, K. E., Stergiou, C., Wang, H., & Gupta, B. B. (2018). Efficient IoT-based sensor BIG Data collection–processing and analysis in smart buildings. *Future Generation Computer Systems*, 82, 349–357.
21. Mamta (2021). "Big Data: The Part and Parcel of Today's Digital World, Insights2Techinfo", p. 1. https://insights2techinfo.com/big-data-the-part-and-parcel-of-todays-digital-world/.

# Chapter 2

# Big Data Management and Modeling

In big data applications, data come in high volume with high velocity and lots of variations. Before dealing with such enormous data, there is a need to know what these data look like. Data modeling and management [1] concepts come into the picture to answer such questions. The purpose of data modeling is to meticulously investigate the nature of data to determine how much storage they will need and what type of processing can be done with them. Data management aims to determine what kind of infrastructure support would be required for the data. Here, the focus will be on different storage architectures for big data. Once all the operational requirements are understood for the data coming from a big data application, one could pick the right system to perform all the required tasks.

## 2.1 Big Data Management

Big data management refers to the effective processing, organization, and use of huge amounts of organized and unstructured data [2,3]. Big data management generally handles tasks ranging from data acquisition to data security [4,5].

### 2.1.1 Data Acquisition/Ingestion

As the name suggests, data acquisition/ingestion involves getting the data into the system. In a large-scale system, data ingestion is not trivial like reading from some file or filling up some web form [6]. Two different scenarios are considered to understand data ingestion in an extensive system. Consider a hospital information system that collects terabytes of medical records from several departments in the first scenario. In this setting, to determine the ingestion infrastructure, there is a need to answer specific questions like how many data sources are there, what their sizes are, whether these sources grow, what the data ingestion rates are, etc. In a hospital information system, there are a variety of data sources like medical images, which are large data objects, and patient records, which are relatively small. The collection of these data is an enormous task. The amount of data contributed by various departmental systems stays relatively constant. The rate of data ingested is proportional to the number of medical activities at that hospital, which is generally not enormous. The next question we need to answer is how to manage erroneous data. Error-handling policies should be defined according to the area of application. For example, even if the data are bad or erroneous in a hospital information system, these can never be discarded because they contain medical records. Generally, a flag is used to point out errors in the data, but it is never removed. These error-handling policies are part of ingestion policies that handle situations like abrupt changes in the data rate. In the second scenario, consider a cloud-based data storage system where people can share media and communicate in real time. In this scenario, the data growth rate is very high due to increasing membership and sharing of content over storage. Here, if the data are erroneous, they can be immediately removed because of the high rate at which these are coming. However, the concerning issue could be related to the handling of the overflow, and for that purpose, the ingestion policy could be to keep the excess data at the site store and ingest only when required. These two scenarios indicate that data ingestion and guidelines on when to consume data are integral to big data systems.

### 2.1.2 Data Storage

The storage module in data management deals with storage-related concerns and how to store big data. In big data storage, primarily, there are two issues that need to be dealt with: capacity and scalability [7]. The capacity issue deals with deciding the memory size, i.e., how much storage should be allocated. The scalability issue deals with determining how to attach the storage to the computer directly or to the network connecting computers in a cluster.

## 2.1.3  Data Quality

After data storage, data quality is the next aspect that needs attention [7]. To ensure data quality, it must be potentially free from errors. But there could be another aspect where we check if the data are helpful for the required task. Now, the question is: Why is there a need to worry about data quality? There are three main reasons to ensure data quality. The first reason focuses on the ultimate use of big data, i.e., to provide actionable insights. Bad data quality produces poor analysis, leading to poor decision-making. The second reason focuses on industries like banking and pharmaceutical industries, where poor-quality financial data or data related to clinical trials can lead to serious legal complications. The third reason focuses on the fact that big data is intended to be used by a third party and should be of high quality to gain trust as a data provider. Recent research by Gartner explains the meaning of "data quality" and how it can be achieved using various data management tools [9].

## 2.1.4  Data Operations

One of the very crucial aspects of data management is to define, implement and test a set of operations that are required for a particular application. There are two broad categories of operations: one works on a single data object, and the other works on a collection of data objects. Performing crop operation, where an image is considered as a single data object, is an example of the first type, while selecting a subset of a set, merging two collections, and computing a function on a collection constitute the examples of the second type. Every operator must be efficient in terms of computation time and memory it takes to perform the required task.

## 2.1.5  Data Scalability

Scalability is often considered in two ways: scaling up (vertical scaling) and scaling out (horizontal scaling). Scaling up refers to adding more processors and RAM to the system, whereas scaling out refers to adding more, but less powerful machines that may interconnect over the network. Out of these two, the scale-out option is increasingly being targeted in the big data environment [10]. The majority of big data management systems in use today are built to work over a network of machines and are able to adapt when new machines are added or when a machine fails.

### 2.1.6 Data Security

When the big data system deals with sensitive information, there arises a need for data security. Moreover, if the big data system is deployed over the cloud, ensuring data security becomes the primary concern, and ensuring data security becomes even more challenging in such an environment [7,8]. In the cloud environment, we need to ensure the security of the machines and the network over which the data are being transferred, however this will significantly increase the overall operational cost. Hence, ensuring security and efficient processing simultaneously is still an open research issue in big data management.

## 2.2  Challenges in Big Data Management: Case Study

To understand the big data management challenges, let us take an example of an energy company called Commonwealth Edison (ComEd) that provides electricity and gas to its customers. In 2015, in a news report [11], the company decided to install 4.7 million smart meters. These smart meters generate real-time data that must be stored and processed at the central facility. According to the report [11], the amount of data generated by these smart meters is 1.5 billion per day. Further, ComEd claimed to offer real-time data to its customers in 15 minutes after consuming and processing the data ingested by their systems. In these 15 minutes, they claimed to give four key computations based on analytics of the data received by their systems. The first is to calculate the consumption pattern for each individual and output a histogram of hourly usage for each user [12]. It needs to be computed daily over larger time intervals to precisely determine the hourly need of the consumers. The second computation determines how much power each user will need depending on the external temperature. This is a statistical problem which often requires fitting a piecewise linear regression model to the data. The third task is to identify everyday consumption patterns that persist regardless of outside weather. This is, again, a statistical problem, which may require a periodic autoregressive model for time-series data. The fourth task is to identify groups of comparable customers based on their use patterns so that it can count the number of different customer groups and create customized energy-saving programs for each group. This calls for comparing similarities across several time-series data, which is challenging. Regardless of the number and complexity of the operations, the company claims to deliver insight in 15 minutes. So, to design a big data system for such a company, there is a need to know how much computations can be executed in parallel, what kind of machines with what capabilities are required to handle the data rate and the number and complexity of analytical operations.

## 2.3  Big Data Modeling

In order to analyze or manipulate data, one needs to determine the characteristics of data, for example, there is data in the form of a record containing three fields, such as employee name, employee ID, and date of birth. This aspect is called the structure of data, and it provides insights into the organization of data. Now, after determining the type, one needs to know the type of operation that can be performed on this data. Finally, the constraints this data should follow must be specified clearly. For example, an organization doesn't take in individuals with age less than 18 years, so the constraint on the data is that today's date minus date of birth should not be less than 18. All of these constitute different parts of understanding the data model. Data models are used to describe the data's characteristics, such as the data's structure, the operations that can be performed on the data, and the constraints the data should follow. In the following sections, these components of the data model are discussed in detail.

### 2.3.1  Data Model Structures

Data modeling is the process of creating structured data inside an information system [13]. The terms "structured data" and "unstructured data" are often used in big data. Let's first understand the meaning of "structure" here by taking an example. Consider a CSV file, as shown in Figure 2.1. This CSV file has three records, each with three fields, often called attributes of data. The first two fields are of type string, and the third is the date field. In this fashion, any number of records can be added, but they all follow the same pattern, i.e., last name, first name, and date of birth. Here, although the content will expand, the structure of the data will stay the same. This recurring data organization pattern makes the file structures.

| Prakash, Ram, 14/03/1988 |
|---|
| Smith, John, 21/08/1970 |
| Singh, Avraj, 28/09/1990 |

**Figure 2.1.**    CSV File 1

| Prakash, Ram, 14/03/1988, Electrical, 70,000 |
|---|
| Smith, John, 21/08/1970, Computers, 80,000 |
| Singh, Avraj, 28/09/1990, Mechanical |
| Steve, Richard, 11/07/1980, 1,50,000 |

**Figure 2.2.**    CSV File 2

$$Structure \begin{cases} A_1, A_2, \ldots, A_k \\ B_1, B_2, \ldots, B_k \\ \ldots \\ \ldots \\ X_1, X_2, \ldots, X_k \end{cases}$$

**Figure 2.3.**    Organizational Structure

Let's consider another CSV File 2, as shown in Figure 2.2. It has four records having five fields: last name, first name, date of birth, branch, and salary. Now, the question is whether this file is structured. If we observe, in the last two records, one of the fields is missing. The missing data make the record incomplete but does not break the structural organization of the data. If we look at Files 1 and 2 together side by side, then File 1 has the same pattern as File 2, consisting of the first three fields. However, if we broadly look at these files, we see that they are both collections of $k$ fields. The difference is in the collection size; File 1 has three records while File 2 has four records. These files share the same data model since it is the same organizational structure which produced them, as shown in Figure 2.3.

In contrast to structured data, unstructured data (Figure 2.4) is quite challenging to interpret and organize.

The unstructured data often include JPEG files, MP3 audio files, and MPEG video files.

```
&#2453;&#2494;&#2480; &#2453;&#2507;&#2469;&#2494;&#2527;&#2469;&#2494;&#2453;&#2494;;
&#2441;&#2458;&#2495;&#2468;&#2476;&#2507;&#2461;&#2494;&#2479;&#2494;&#2458;&#2509;&#2459;&#2503;;
&#2472;&#2494; &#2439;&#2470;&#2494;&#24#2472;&#2496;&#2434;;! &#2456;&#2480;&#2503;;
&#2469;&#2494;&#2453;&#2476;&#2503; &#2453;&#2503;;,
&#2438;&#2480;&#2476;&#2494;&#2439;&#2480;&#2503;&#2439; &#2476;&#2494; &#2453;&#2503;;,
&#2476;&#2480;&#2509;&#2471;&#2478;&#2494;&#2472;&#2503; &#2453;&#2494;&#2480;;
&#2469;&#2494;&#2453;&#2494; &#2470;&#2480;&#2453;&#2494;&#2480;;, &#2453;&#2494;&#2480;;
&#2458;&#2482;&#2503; &#2479;&#2494;&#2451;&#2527;&#2494; &#2470;&#2480;&#2453;&#2494;&#2480;;
&#2478;&#2494;&#2482;&#2470;&#2489; &#2469;&#2503;&#2453;&#2503;&mdash; &#2488;&#2476;;
&#2453;&#2503;&#2478;&#2472; &#2455;&#2497;&#2482;&#2495;&#2527;&#2503;;
```

**Figure 2.4.**   Unstructured Data

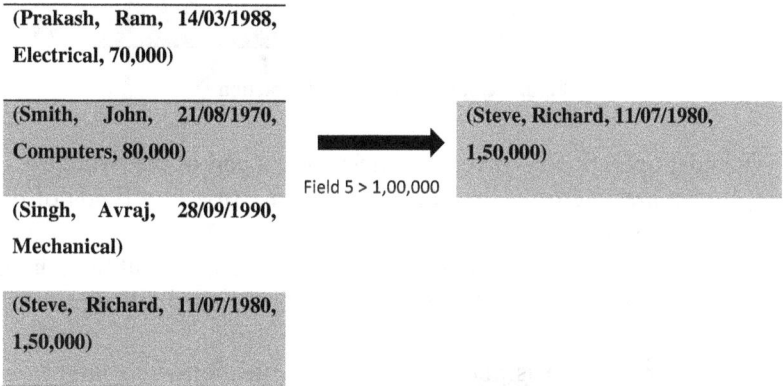

(Prakash, Ram, 14/03/1988, Electrical, 70,000)

(Smith, John, 21/08/1970, Computers, 80,000)

(Singh, Avraj, 28/09/1990, Mechanical)

(Steve, Richard, 11/07/1980, 1,50,000)

Field 5 > 1,00,000

(Steve, Richard, 11/07/1980, 1,50,000)

**Figure 2.5.**   Subset Operation

## 2.3.2 Data Model Operations

The second component of a data model is a set of operations that can be performed on the data. Data operations specify the methods for the manipulation of data. Different data models are associated with different structures, so the operations are also different. But certain operations are performed across all the data models. The subset, substructure extraction, union, and join are some of the operations [14].

In subset operation, a part of the collection is extracted based on certain conditions. For example, there is a collection of records, as shown in Figure 2.5, and we want a subset of the records in a collection that meets the requirement that the fifth field has a value larger than 1,00,000. This operation is also called selection or filtration.

| (Prakash, Ram, 14/03/1988, Electrical, 70,000) | (Prakash, Ram) |
| (Smith, John, 21/08/1970, Computers, 80,000) | (Steve, Richard) |
| (Singh, Avraj, 28/09/1990, Mechanical) | (Singh, Avraj) |
| (Steve, Richard, 11/07/1980, 1,50,000) | (Steve, Richard) |

Field 1 & Field 2

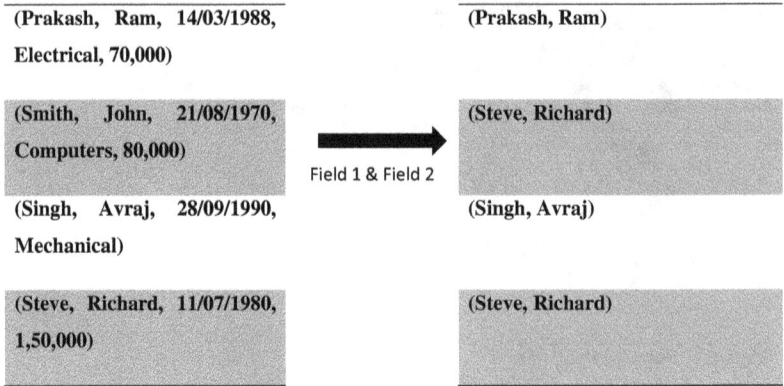

**Figure 2.6.**   Substructure Extraction

The next operation involves the retrieval of a part of the structure. For example, we are interested in extracting just the first two fields of each record, as shown in Figure 2.6.

This operation results in a new collection of records with just the last name and first name fields. This operation is dominantly called projection operation.

The next operation is about combining collections into larger ones. There are several ways to interpret the word "combine". The union is the simplest of them all. The union operation is based on the assumption that the two collections must have identical structures. In other words, two collections have different numbers of fields. For example, one collection could have four fields while the other might have 14, and these cannot be combined using union.

The union operation combines two collections to make a new collection and eliminate duplicates, as shown in Figure 2.7. There is a generalization of this concept which allows duplicates and produces five records in the output. Another kind of combining collections is called join. The join operation can be performed when two collections have distinct data contents but share certain items. There are two stages in combining data using join. In the first stage, look for the common field in both collections, locate a set of matching data items in collection 2, and put all fields of the matching record pairs, as shown in Figure 2.8.

Typically, there would be many record pairings that matched each other, thus making the operation complex and expensive as data size increases.

(Prakash, Ram, 14/03/1988)

(Smith, John, 21/08/1970)

(Richard, Steve, 11/07/1980)

———————————————
*Collection 1*

(Singh, Avraj, 28/09/1990)

(Richard, Steve, 11/07/1980)

———————————————
*Collection 2*

UNION

*(Prakash, Ram, 14/03/1988)*

*(Smith, John, 21/08/1970)*

*(Richard, Steve, 11/07/1980)*

*(Singh, Avraj, 28/09/1990)*

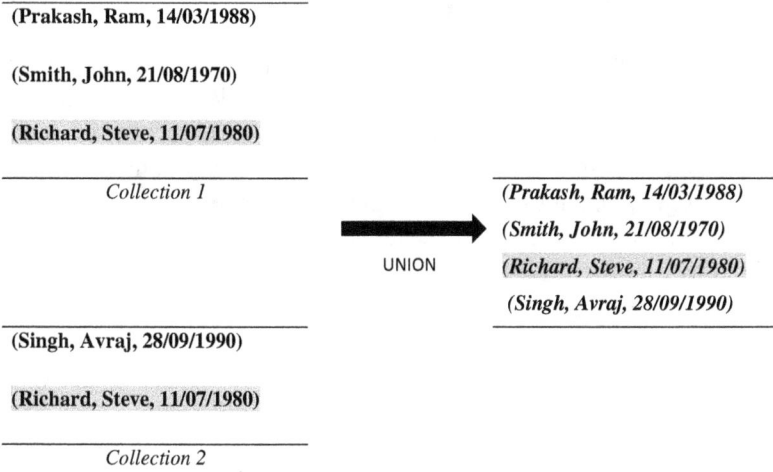

**Figure 2.7.**   Union of Collections 1 and 2

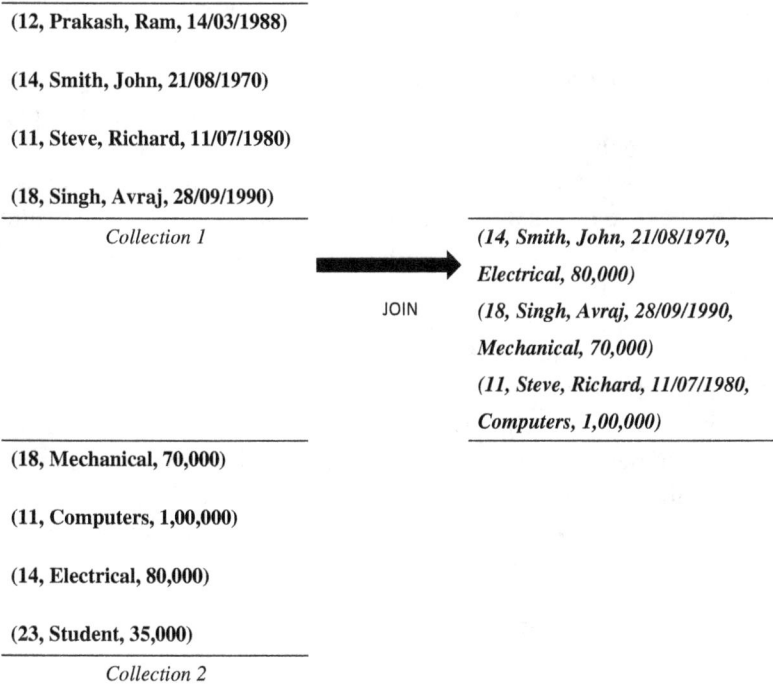

(12, Prakash, Ram, 14/03/1988)

(14, Smith, John, 21/08/1970)

(11, Steve, Richard, 11/07/1980)

(18, Singh, Avraj, 28/09/1990)

———————————————
*Collection 1*

(18, Mechanical, 70,000)

(11, Computers, 1,00,000)

(14, Electrical, 80,000)

(23, Student, 35,000)

———————————————
*Collection 2*

JOIN

*(14, Smith, John, 21/08/1970, Electrical, 80,000)*

*(18, Singh, Avraj, 28/09/1990, Mechanical, 70,000)*

*(11, Steve, Richard, 11/07/1980, Computers, 1,00,000)*

**Figure 2.8.**   Join Operation of Collections 1 and 2

### 2.3.3 Data Model Constraints

The third component of the data model is a constraint. A constraint is a logical statement, implying that it is possible to calculate and verify that the assertion is accurate. Since they can define some aspects of the semantics or meaning of the data, constraints are a necessary component of the data model [15]. A data system, for instance, would not be aware of the limitation that a week contains seven days unless the information was provided to it in the form of a constraint. Another example of constraint may be that only a single title is allowed in movies, which is a type of uniqueness constraint. Constraints can take many different forms. Different types of constraints will apply to various data models. A value constraint is an inference about the values of specific data. For example, the value constraint prohibits the value of data components reflecting the age from being negative. Another type of constraint is the cardinality constraint of data property. It needs us to count the number of values each object has connected with it and determine whether or not it falls inside an upper and a lower bound. For example, a person can take sugar medications between 0 and 3. A new form of value constraint may be applied by limiting the data types permitted in a field. For example, a person's name is a non-numeric alphabet string, and we can set the type constraint as (Name: string, not(isNumeric(Name))). The type constraint is a special type of domain constraint. The range of values assigned to a data property or attribute is known as its domain. For instance, the range of permissible values for the date field's day component is between 1 and 31. In contrast, the value of a month might range from 1 to 12; alternately, the value of a month may range beginning with January and ending with December. All these are variations of the value constraint, which states how to restrict the values of some data properties. Contrarily, structural constraints place limitations on the data's structural characteristics rather than the values. For example, we have a matrix that is restricted to a square matrix in which the numbers of rows and columns are equal. There is no restriction on the numbers of rows and columns; the only restriction is that they should be equal. Now that the matrix structure is constrained, the total number of entries will be squared.

## 2.4 Types of Data Models

As discussed in the previous section, three components characterize a data model: the structure of data, the operations on that structure, and the constraints the data should follow. This section discusses various commonly used data models.

## 2.4.1 Relational Data Model

It is one of the most basic and widely used data models today. It is a structured data model which is the foundation for several well-known conventional database management systems, including MySQL, Oracle, Teradata, and others. The primary data structure for a relational model is a table similar to the one shown in Table 2.1.

Table 2.1 represents a set of tuples where each row represents a tuple, which was earlier referred to as a record. The elements of each tuple are atomic, i.e., these cannot be further divided. Each element in a tuple represents one unit of information. Table 2.1 illustrates the relation of five tuples. By the definition of a set, which is the collection of distinct objects, we cannot add the following tuple, (106, Alice Cooper, IT, Security Specialist, 1,457,800), because that will introduce a duplicate. Another tuple, (Jane, Doe, 109, Research Associate, 1,308,900), also cannot be added here because it is in a different order. Now, the question is how does the system recognize that this tuple is different? The system recognizes this by looking at the schema, which is given in the header of Table 2.1. The schema gives us information about the name of the table and the name of the attributes. It further gives information about the data types that are permitted for each attribute and the constraints that should be satisfied by each column. Given this schema, now the system knows that the above-defined two tuples are invalid and are not allowed.

**Table 2.1.** A Typical Relational Model Employee

| ID: Int (Primary Key) | FName: String (Not Null) | LName: String (Not Null) | Department: String | Title: String | Salary: Int > 1,000,000 |
|---|---|---|---|---|---|
| 102 | Ram | Prakash | IT | DBA | 1,050,000 |
| 103 | John | Smith | IT | Programmer | 1,175,000 |
| 104 | Steve | Richard | Research | Director | 2,056,000 |
| 105 | Avraj | Singh | HR | CA | 9,60,000 |
| 106 | Alice | Cooper | IT | Security Specialist | 1,457,800 |

## 2.4.2  Semi-Structured Data Model

The largest information source available now is the internet, and the data model that underlies the web is the semi-structured data model [16]. Let's take an example of a very simple web page without much styling or linking to other web pages, as shown in Figure 2.9. The Hyper Text Markup Language (HTML) code shown in Figure 2.9 is used by the web browser to render this web page. Here, every block is nested inside the larger block and is unlike the structured relational model. Further, a document can have a varying number of them. So, there is some structure, but it is quite flexible as compared to the relational model.

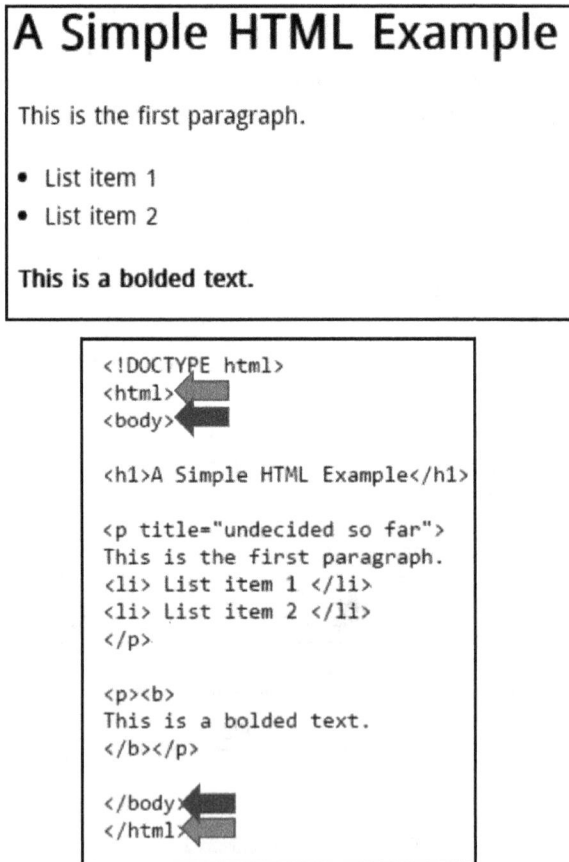

# A Simple HTML Example

This is the first paragraph.

• List item 1
• List item 2

**This is a bolded text.**

```
<!DOCTYPE html>
<html>
<body>

<h1>A Simple HTML Example</h1>

<p title="undecided so far">
This is the first paragraph.
<li> List item 1 </li>
<li> List item 2 </li>
</p>

<p><b>
This is a bolded text.
</b></p>

</body>
</html>
```

**Figure 2.9.**   Web Page Example and the Associated HTML Code

Extensible markup language (XML) is another widely used standard for representing data onto the web [14,15]. It is the generalization of HTML and is even more flexible than HTML. The markers can be any string unlike HTML where we have standard markers. Another very popular data format especially in the context of big data is Java Script Object Notation (JSON). It is widely used on the platforms like Twitter, Facebook etc. A sample JSON file is shown in Figure 2.10. Here, we can see nested structures. The atomic component in JSON file is the key, value pair.

All these different types of semi-structured data formats can be generalized by modeling them as trees. Let's look at the XML document and its corresponding tree structure, as shown in Figure 2.11.

```
{
    "status": 200,
    "photos":
    [
        {
            "typeName": "Facebook",
            "type": "facebook",
            "typeId": "facebook",
            "url": "http://graph.facebook.com/amoghnatu/picture?type=large",
            "isPrimary": true
        }
    ],
    "contactInfo": {
        "familyName": "Natu",
        "fullName": "Amogh Natu",
        "givenName": "Amogh"
    },
    "demographics": {
        "gender": "male"
    },
    "socialProfiles":
    [
        {
            "id": "1839143973",
            "typeName": "Facebook",
            "username": "amoghnatu",
            "type": "facebook",
            "typeId": "facebook",
            "url": "http://www.facebook.com/amoghnatu",
        }
    ]
}
```

**Figure 2.10.** JSON File

```
<document>
  <report>
    <author>Video database</author >
    <date>June 12, 2000</date>
  </report >
  <paper>
    <title>XML query data model</title>
    <author>Don Robie</author>
    <source>W3C, June 2000</source>
  </paper>
</document>
```

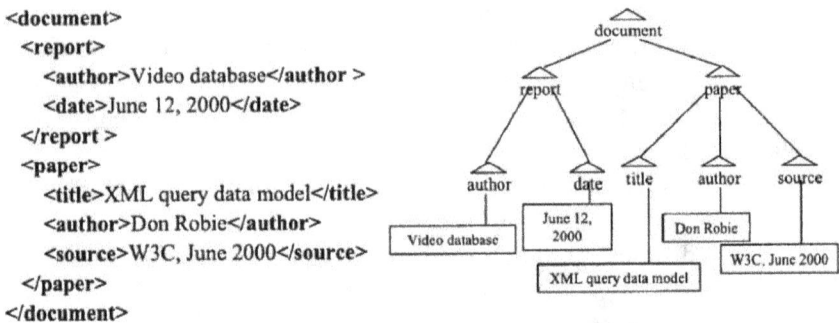

**Figure 2.11.**   XML Data Model with Associated Tree Structure

As can be seen in Figure 2.11, the document represents the most important item in the XML file, which also makes it the root of the tree. We now have a report element with an author and date, as well as a paper element with a title, author, and source. Both of these components are located under the document category. The actual values, such as "Don Robie", are considered to be the textual content. These text values are always considered to be the leaves of the tree because a text item can never have any more elements added to it. Now, there are several benefits to representing a document as a tree because it is a well-known data structure that enables navigational access to data. The operations like getParent, getChild, and getSiblings help in navigating the document with great ease. Textual inquiries are another option for us, and we could ask something like, "Who wrote the XML query data model?"

### 2.4.3  Unstructured Data Model: Vector Space Data Model

To extract information from the big text and picture collections, a vector space model is utilized. Text is almost always assumed to be an example of unstructured data because it does not contain any attributes or relationships. In its place, it merely consists of a series of strings that are broken up by punctuation. In order to find a text from a large collection and to analyze the text data, one needs to look at different ways. The data models for structured and unstructured data will not apply here. In order to find out the text data, we also require a separate structure called vector model [16] that is generated from the text data in addition to the text data itself. Moreover, discovering a document is not really an issue that involves an exact search. First, the system is provided with a query document, and then we ask it to locate any documents that are similar to it. The question that arises now is how this similarity is estimated and applied to the document-searching process. To answer all these questions, let's consider three types of documents and the term frequency of each word in the document in the form of a matrix, as shown below in Table 2.2.

From the term frequency (TF) matrix, a new vector is created for each term which is called as the inverse document frequency (IDF). Table 2.3 shows how these IDF values are computed.

In the calculation of IDF, why have we used $\log_2$ instead of $\log_{10}$? There is not much technical reason behind this, but it is used as a convention in computer science because many numbers are in the power of 2. Further, $\log_2 x = (\log_{10} x)(\log_2 x)$, where $\log_2 10$ is a constant, hence the relative score of IDF does not change regardless of the base we use. The document frequency for each term is calculated by dividing its count over the whole collection by the total number of documents. To find IDF, we take the inverse of the document frequency, so that $n$, the number of documents, is in the numerator. To understand the intuition behind IDF, suppose we have 100 random newspaper articles and let's say 10 of them cover elections. Let's say that in these 10 articles, the term "election" occurs 50 times and the verb "is" occurs, let's say, 300 times because it is

**Table 2.2.** Documents with the Term Frequency Matrix

|     | Hindu | Hindustan | Times | India |
| --- | --- | --- | --- | --- |
| d1  | 1 | 0 | 0 | 0 |
| d2  | 0 | 1 | 1 | 0 |
| d3  | 0 | 0 | 1 | 1 |

*Notes*: d1: The Hindu; d2: Hindustan Times; d3: Times of India.

**Table 2.3.** Inverse Document Frequency

| Term | Term Frequency | Document Frequency | Inverse Document Frequency |
| --- | --- | --- | --- |
| Hindu | 1 | 1/3 | $\log_2\left(\frac{3}{1}\right) = 1.584$ |
| Hindustan | 1 | 1/3 | $\log_2\left(\frac{3}{1}\right) = 1.584$ |
| Times | 2 | 2/3 | $\log_2\left(\frac{3}{2}\right) = 0.584$ |
| India | 1 | 1/3 | $\log_2\left(\frac{3}{1}\right) = 1.584$ |

**Table 2.4.**  TF–IDF Matrix

|     | Hindu | Hindustan | Times | India | Length |
|-----|-------|-----------|-------|-------|--------|
| d1  | 1.584 | 0         | 0     | 0     | 1.584  |
| d2  | 0     | 1.584     | 0.584 | 0     | 1.688  |
| d3  | 0     | 0         | 0.584 | 1.584 | 1.688  |

the most frequently occurring term, which is also called the stop word. If we compare the document frequency of both these terms, the document frequency of "is" is six times the document frequency of the other term "election". But it seems wrong because "is" is such a common word that its prevalence has a negative impact on its informativeness. So, now, if you want to compute the IDF of "is" and "election", the IDF of "is" will be far lower. So, IDF acts like a penalty factor for terms which are too widely used to be considered informative. As we know IDF is a penalty factor, if we multiply the TF with IDF, we get the actual importance of a term in the document, which is also called the rank. The columnwise multiplication of TF with IDF gives us the TF–IDF matrix, as shown in Table 2.4.

So, for each document, we have a vector which is represented as a row, as shown in Table 2.4. Here, a row represents the relative importance of each term in the vocabulary. Vocabulary is the collection of all words that appear in this collection. If the vocabulary has 3 million entries, then this vector can get quite long. Additionally, if the number of documents increases to, say, 1 billion, it becomes a big data problem. The last column in Table 2.4 represents the length of the document vector. It is the square root of the sum of squares of the individual term scores, as given by the following formula:

$$\text{Length of d2} = \sqrt{1.584^2 + 0.584^2} = 1.688 \tag{2.1}$$

Now, to perform a search in the vector space, let's say, we write a query, *q: Times Hindustan Times*. In the query, the number of terms is three, and out of these, the term "Times" appears twice. This is in fact the maximum frequency out of all the terms in the query. Now, to create a query vector, we take the document vector of the query and multiply each term by the number of occurrences divided by two which is the maximum term frequency. The resultant query vector is $Q = \left[0, \frac{1}{2} \times 1.584, \frac{2}{2} \times 0.584, 0, 0\right]$.

Now, we can calculate the length of the query vector as, $\sqrt{0.792^2 + 0.584^2} = 0.984$. Next, we have to find the similarity between the query vector and each document, i.e., how far the query vector is from each document.

There are several similarity functions established nowadays, which are employed for various purposes. The cosine function, which calculates the cosine function of the angle between these two vectors, is a widely used similarity metric. Equation (2.2) gives the formula for computing the function:

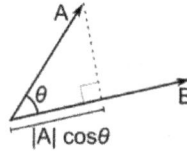

$$\text{Similarity} = \cos(\theta) = \frac{A \cdot B}{\|A\|\|B\|} = \frac{\sum_{i=1}^{n} A_i B_i}{\sqrt{\sum_{i=1}^{n} A_i^2}\sqrt{\sum_{i=1}^{n} B_i^2}} \qquad (2.2)$$

The angle between two vectors should be 0 if they are equal, and thus, the cosine function consequently evaluates to one. As the angle increases, the value of the cosine function decreases to make them more dissimilar. Using the above formula, the similarity of the query vector with each document vector is computed as follows:

Similarity: d1[1.584,0,0,0] and Q[0,0.792,0.584,0]:

$$\cos\text{Sim}(d1,q) = \frac{1.584 \times 0 + 0 \times 0.792 + 0 \times 0.584 + 0 \times 0}{1.584 \times 0.984} = 0 \quad (2.3)$$

Similarity: d2[0,1.584,0.584,0] and Q[0,0.792,0.584,0]:

$$\cos\text{Sim}(d2,q) = \frac{0 \times 0 + 1.584 \times 0.792 + 0.584 \times 0.584 + 0 \times 0}{1.688 \times 0.984} = 0.96$$

$$(2.4)$$

Similarity: d3[0,0,0.584,1.584] and Q[0,0.792,0.584,0]:

$$\cos\text{Sim}(d2,q) = \frac{0 \times 0 + 0 \times 0.792 + 0.584 \times 0.584 + 1.584 \times 0}{1.688 \times 0.984} = 0.21$$

$$(2.5)$$

If we examine the distance function's output, document 2 (d2) is far closer to the query than the other two.

Nowadays, the vector space model is also used for similarity search in images. One can compute features from images, and one common feature is a scatter histogram. Further, the texture of the image, the forms of the objects, and any other features may be calculated as similarity vectors. Thus, making a vector space model is significant for unstructured data.

### 2.4.4  Graph Data Model

The next category of data has the form of graphs or networks, and the most obvious examples are social networks. The Lord of the Rings Trilogy inspired Tim Libzek to establish a social network, as shown in Figure 2.12 [17]. This graph illustrates the characters' loyalties.

The nodes in the graph are characters and other entities, like cities, and the edges connecting pairs of nodes represent allegiances.

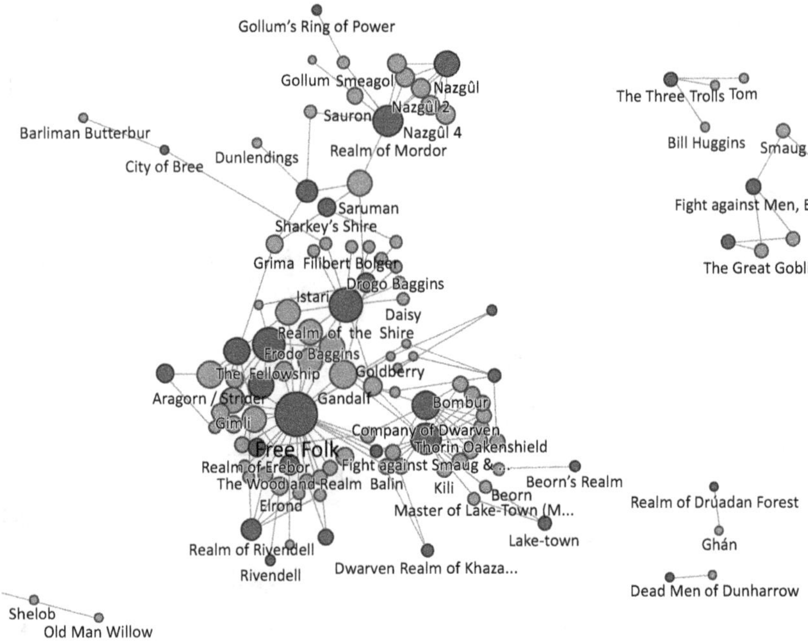

**Figure 2.12.**    Social Network for Lord of Rings Trilogy [17]

The fact that a graph contains two different types of information sets it apart from other data models: (1) properties and attributes of entities and relationships and (2) the connectivity structure that constitutes the network itself [18]. To understand the data and connection part of a graph model, consider a property graph, as shown in Figure 2.13, taken from the Apache Spark system [19].

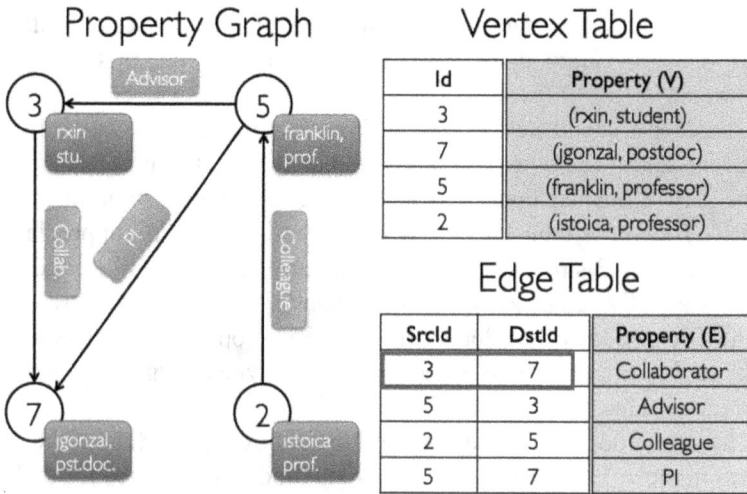

**Figure 2.13.** Property Graph from Apache Spark [18]

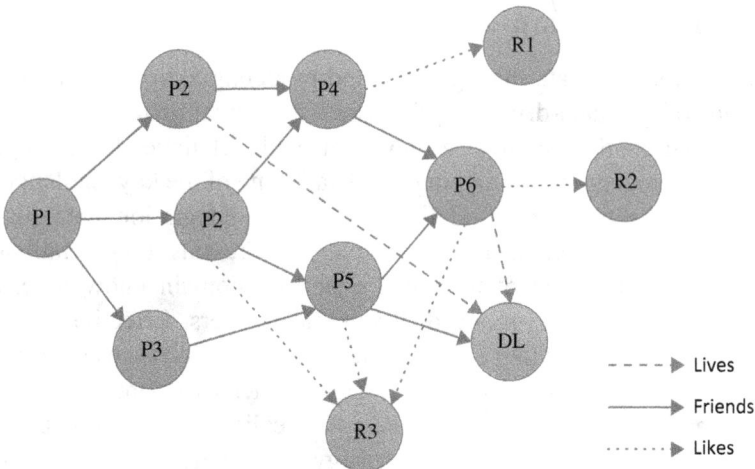

**Figure 2.14.** A Sample Social Network

This graph is represented by the vertex table and the edge table. The vertex, or node table, gives IDs to nodes and lists their properties. The edge table has two parts: the colored part represents the properties of the edge, whereas the white part contains just the direction of the arrows in the network.

As the connection information is included, it gives graph data a different sort of computational capability than previous data models we have examined so far. Even without examining the characteristics of nodes and edges, one may learn a lot by examining or querying this connection structure.

Consider a social network with three kinds of nodes, user, city, and restaurant, and three types of edges, friends, likes, and lives, as shown in Figure 2.14. Consider that P1 is looking for a decent restaurant in Delhi that P1's friends or their friends, who also reside in Delhi, appreciate. P1 may decide on R3 since it receives the greatest amount of like edges from residents who live close to Delhi. Additionally, it may be found by following the friend edges that depart from P1. The operation that is used to arrive at this conclusion is called traversal, where edges are followed based on certain conditions.

The data models discussed in Section 2.4 cover just a few of the popular data models. There are many other data models developed for different purposes and application scenarios.

## 2.5  Chapter Summary

In conclusion, big data management and modeling are essential components in today's data-driven world. The exponential growth of data has necessitated the development of new tools and techniques for managing, storing, processing, and analyzing big data. Some of the key challenges in big data management and modeling include data integration, data quality, data security, scalability, and complexity. Addressing these challenges requires a combination of technical expertise, domain knowledge, and effective collaboration among different stakeholders. Effective big data management and modeling can provide several benefits, such as improved decision-making, better customer insights, increased operational efficiency, and reduced costs. However, the realization of these benefits requires a well-designed big data strategy that aligns with the organization's goals and objectives.

# References

1. Chen, P. P. (1976). The entity-relationship model: Toward a unified view of data. *ACM Transactions on Database Systems (TODS)*, 1(1), 9–36.
2. Manyika, J., Chui, M., Brown, B., Bughin, J., Dobbs, R., Roxburgh, C., & Byers, A. H. (2011). *Big Data: The Next Frontier for Innovation, Competition, and Productivity*. McKinsey Global Institute, Seattle.
3. Stergiou, C. L., Psannis, K. E., & Gupta, B. B. (2021). InFeMo: Flexible big data management through a federated cloud system. *ACM Transactions on Internet Technology (TOIT)*, 22(2), 1–22.
4. Sharma A. & Chhabra A., (2022) Big data: The Future of information management. In *Data Science Insights Magazine* (Vol. 1, pp. 21–24), Insights2 Techinfo.
5. Gupta A. K. & Chui K. T. (2022), Blockchain-based decentralized data management for the metaverse. In *Data Science Insights Magazine* (Vol. 3, pp. 21–24), Insights2Techinfo.
6. Gou, L. & Zhang, W. (2012). Research on the acquisition and management of big data. In 2012 *IEEE International Conference on Computer Science and Automation Engineering (CSAE)* (pp. 684–687).
7. Kaisler, S., Armour, F., Espinosa, J. A., & Money, W. (2013). Big data: Issues and challenges moving forward. *Proceedings of the 46th Hawaii International Conference on System Sciences* (pp. 995–1004).
8. Gupta A. K. & Santaniello D. (2022). Performance analysis of big data and blockchain based IoT security techniques. In *Data Science Insights Magazine* (Vol. 1, pp. 15–20), Insights2Techinfo.
9. Jason Medd, "Data Quality Fundamentals for Data and Analytics Technical Professionals", https://www.gartner.com/en/documents/4014199.
10. Zhang, X., Cheng, X., & Hu, Y. (2015). Big data: Understanding big data. *Journal of Signal Processing Systems*, 80(3), 213–228.
11. Katherine Tweed (2015). "New York Prepares for Millions of Smart Meters Under REV", https://www.greentechmedia.com/articles/read/new-york-prepares-for-millions-of-smart-meters-under-rev.
12. Sharma, A., Singh, S. K., Badwal, E., Kumar, S., Gupta, B. B., Arya, V., ... & Santaniello, D. (2023, January). Fuzzy based clustering of consumers' big data in industrial applications. In *2023 IEEE International Conference on Consumer Electronics (ICCE)* (pp. 1–3).
13. Fayyad, U., Piatetsky-Shapiro, G., & Smyth, P. (1996). From data mining to knowledge discovery in databases. *AI Magazine*, 17(3), 37–54.
14. Silberschatz, A., Korth, H. F., & Sudarshan, S. (2010). *Database System Concepts*. McGraw-Hill, New York.
15. Halpin, T. & Morgan, T. (2008). *Information Modeling and Relational Databases*. Morgan Kaufmann, Elsevier, USA.

16. Date, C. J. (2004). *An Introduction to Database Systems*. Addison-Wesley, Pearson Education, USA.

17. Orrego, D. L. & Casas-Roma, J. (2018). Social network analysis of The Lord of the Rings: The structure of J.R.R. Tolkien's epic. *Social Network Analysis and Mining*, 8(1), 11.

18. Angles, R. & Gutierrez, C. (2008). Survey of graph database models. *ACM Computing Surveys (CSUR)*, 40(1), 1–39.

19. Xu, H., Hu, Y., Lin, W., Zhang, J., Wang, Y., & Ye, Y. (2019). A hybrid graph storage for property graph in Apache Spark. In *Advances in Big Data and Cloud Computing* (pp. 125–136), Springer, Singapore.

# Chapter 3

# Big Data Processing

This chapter focuses on the issues of ingesting and processing big data and also throws light on how the processing of big data is influenced by the data's variety, volume, and velocity.

## 3.1 Requirements for Big Data Processing

Consider a use case of online gaming to understand the requirements of big data processing and how it differs from processing in traditional data architecture. When a game is just launched and users start the registration process, one can begin with the traditional relational database to track user sessions and other relevant information. The game server receives an event alert whenever a player starts a session and scores a point in the game. The database can initially manage such event streams entering the server. However, suppose the game unexpectedly gets popular. In this case, the game server's database management system won't be able to handle the demand because the rate at which events are arriving makes it impossible to enter them into the database. A buffer can be used to process this growing data to address this issue. However, as the demand for the game goes up with time, more processing nodes (buffer) and even more database servers will be needed to handle the load. This is a typical situation that most websites encounter when dealing with big data problems linked to information volume and velocity. This example shows that the issue can be solved in a single step initially. But the system becomes less reliable and more difficult to evolve as the game creators include more reactive patches. To keep up with the load on the system, the game

43

creators need to address various infrastructure management-related challenges. However, the database servers are also susceptible to damage and corruption, which need to be managed individually in terms of replication and fault tolerance.

Let's begin by going over these concerns. Imagine if one of the processing nodes failed. There may be some data loss in the meantime as the system manages and restarts the procedure. Before data could be discarded, the system would need to inspect each processing node. It is necessary to replicate every node and every database independently. Further, batch calculations that need data from many data servers must access and keep up with the data independently, which might be time-consuming and expensive. Managing the aforementioned complications, such as failed servers and malfunctioning processing nodes, will be easier with big data processing tools like the Spark system.

To conclude, the main requirements and expectations of this gaming application from a big data system are as follows: (i) To handle and divide enormous amounts of event data streaming into the system, there must be a mechanism to leverage basic big data techniques. Specifically, this refers to the division and placement of data in and out of computer memory along with a model for subsequent synchronization of the datasets. (ii) The data access should be fast. (iii) The system should possess the ability to simultaneously deploy several event processing jobs to distributed processing nodes. (iv) Scaling out should also be possible, i.e., adding extra resources, such as distributed computers, to scale up and process faster without sacrificing performance. (v) It needs to offer computational reliability and fault tolerance in the event of malfunctions. This entails providing programmable replications and event data recovery as necessary. (vi) It should handle various data types like text-based chats, images, scores, and graphs of players. (vii) It should be able to perform both streaming and batch processing. It should be able to handle operations at small chunks of data streams with minimal delay, which means the system should have low latency. At the same time, it should handle processing all available data in batch form. Latency is the term that is often heard and used in big data processing. Latency quantifies the delay in processing the data streaming into the system, i.e., the difference between production or event time and processing time of a data entry.

Finally, the actual reason for all these big data processing requirements, which differ from those for processing in typical data architecture, is that big data has variable volume and velocity, necessitating dynamic and scalable batch and stream processing. Big data involves various requirements, including managing and integrating data across several data systems.

## 3.2 Big Data Retrieval

The process of specifying and obtaining requested data from data storage is known as data retrieval [1]. Assume that the data is kept in a repository that adheres to a specific data model, such as the relational data model. In this context, data retrieval is used in two ways. The first is specifying a data request for static and streaming data called the query specification method. The second component of the data retrieval process is the internal mechanism, which refers to the processing that occurs within the data management system to either compute or assess the specific retrieval request. In spite of the fact that query specification can be utilized for both small and large data stores, a significant amount of focus will be placed on big data systems. The focus will be on the evaluation of queries when applied to large amounts of data. In addition, we will concentrate on the ways in which the query specification shifts when we are working with faster streaming data.

### 3.2.1 Relational Data Query

A set of relations make up a relational database. A relation consists of two components: (i) a table with rows and columns; (ii) a schema that specifies the name of the relation as well as the name and type of each column. A series of tables and columns are used to arrange the items. Each column in a table contains a specific type of data, whereas a row in a table represents a collection of related values of one object [2]. A major strength of the relational models is that they support simple, powerful data querying. A relational database query is a question about the data, and the answer consists of a relation containing the result. The language used to specify a query request is known as query language. Declarative query languages allow us to specify the data we wish to get without instructing the machine on how to do so. For instance, we want to find all employee's data whose salary is greater than $50,000. To retrieve this data, there is no need to create a program that instructs the operating system to open a file, read an integer, and other such details. In this step, the system takes care of everything after we have specified the necessary data items. Structured query language, also known as SQL, is the most popular query language to use while working with relational data. In contrast to a query language, a database programming language is a high-level procedural programming language that includes embedded query operations. Examples of database programming languages include Oracle's PL/SQL and Postgres's PgSQL. In situations involving structured data, the query language of choice is SQL. Both traditional database management systems like Oracle and modern distributed systems like Hadoop Spark use it. SQL is the standard for querying structured data, which ranges from Oracle SQL to Spark SQL. We will now use an example as an illustration. We must first specify the database schema. Imagine a company named the Organic India Club that operates several franchises across India where organic beverages/drinks are sold. The schema of this business has six relational tables: Franchises(name, address, license), Drinks(name, manufacturer), Sells(franchise, drink, price), Drinkers(name, address, phone), Frequents(drinker, franchise), Likes(drinker, drink). The attributes that are underlined represent the primary key. The primary key refers to attributes that make a record unique. The names, addresses, and license numbers of the franchises are all included in the first table. The second table, titled Drinks, contains the names of various organic beverages along with the manufacturers who produce them. Not every franchise offers the same kinds of beverages, and even when they do, variable establishment expenses may result in varied prices for the same product. Therefore, the Sells table keeps track of which franchise offers which drink at what cost. This company maintains the information about its regular member customers, which is one of the interesting aspects of the company. Therefore, the names, addresses, and contact information for these customers are listed in the Drinker table. Furthermore, it knows which franchise members visit and which drinks they prefer.

A SELECT-FROM-WHERE clause is the most fundamental building block of an SQL query's structure. For example, we want to retrieve the names of the beverages sold by Organico Agro. The following query shown in Equation (3.1) helps in accomplishing this task:

SELECT name                                   # Output attribute(s)
FROM Drinks                                    # Table(s) to use            (3.1)
WHERE manufacturer = 'Organico Agro'  # Condition(s) to satisfy

The results of the query are presented in the form of a table with a single column labelled name.

Now, consider the scenario that the table "drinks" is large and has millions of entries. In such a scenario, we need to divide/partition the table across several machines. This can be done because the query has to SELECT and PROJECT, which can be done in parallel. As we know, "name" is the table's primary key. The method of partitioning known as range partitioning by the primary key is one of the standard approaches. It indicates that the rows of the table are organized into groups according to the alphabetical order of the name value, as demonstrated in Table 3.1. All database management companies, including IBM, Microsoft, and others, have a solution similar to this for dealing with large volumes of data, and it involves the use of data partitioning. However, newer systems such as Spark are inherently distributed and thus offer partitioning as part of their structure. Now, the task is to perform queries over the partitioned tables. Let's look at the two queries for such partitioned tables. First, find records for the drinks whose name starts with 'Co', and second, ask for the names of the beverages sold by Organico Agro like before.

The two queries are shown as follows in Equation (3.2):

**Query 1:**
SELECT *                                      # Output attribute(s)
FROM Drinks                                   # Table(s) to use
WHERE name like 'Co%'                         # Condition(s) to satisfy

**Query 2:**                                                                      (3.2)
SELECT name                                   # Output attribute(s)
FROM Drinks                                    # Table(s) to use
WHERE manufacturer = 'Organico Agro'  # Condition(s) to satisfy

**Table 3.1.** Large Table Partitioning to Different Machines

| Name | Manufacturer | Name | Manufacturer | | Name | Manufacturer | |
|------|--------------|------|--------------|---|------|--------------|---|
| A | Organico Agro | C | Shivalik | | K | Moriko | |
| A | Moriko | C | VedShakti | ... | K | Organico Agro | ... |
| ... | | ... | | | ... | | |
| B | Rus Organic | D | Saukhya | | L | Sovam | |
| ... | | ... | | | ... | | |

| Machine 1 | Machine 2 | Machine 5 |
|-----------|-----------|-----------|

There are now two additional syntax elements in query 1. The first is a predicate known as "like", and when we use it, we inform the query engine that we merely have partial information about the string that we want to match. The query engine refers to the string that has only been partially specified as a string pattern. The part of the string that we know is written as 'Co', and the part that we don't know is written as '%'. The second query is the same as the one that was discussed earlier; however, this is more difficult to accomplish in a partitioned database, which is typically encountered when working with large amounts of data. Let us take a look at the very first query that was run in this data partitioning setting. Now, the question that needs to be answered is whether or not it is necessary to touch each partition in order to get the results of the first query. The answer is "no". As is common knowledge, the name column serves as the primary key for the table titled "drinks". In addition to this, we are aware that the system actually did plan partitioning based on the name attribute. This indicates that the process of evaluation should only access Machine 2 because no other machine is likely to have records for names beginning with the letter C. Therefore, the system can work considerably more effectively if it is aware of the partitioning scheme. This kind of performance is crucial when a system handles thousands of queries per second. In query 2, the query condition is on the second attribute: manufacturer. This time, the query condition is applied to a different attribute. Thus, we cannot leverage the information from the partitioning to our advantage. Therefore, this query must be sent to all partitions, or in technical terms, it must be broadcasted from the primary machine to all the machines. This broadcast query will be executed on each local machine independently and concurrently. The output from each local machine is brought back into the primary machine, where these are combined together and finally shown to the client. Although it may look like a lot of work is being done, everything is actually being done in parallel, which is the most effective way to deal with large amounts of data. Now, you could be asking yourself, "What if I have 100 machines, but only 20 of them contain the needed data?" Should we exhaustively search all 100 machines just to discover nothing in 80 of them and have those machines provide 0 results? Why then put forth the extra effort? Is it unavoidable? In order to achieve this, an additional component is required in the solution, known as an index structure. If you have an index and a reverse table, you can use the index to get all the entries in the table that contain a given value by specifying the column in which that value appears, as shown in Table 3.2.

**Table 3.2.** Local and Global Index Tables for Manufacturers

| Local | |
|---|---|
| Manufacturers | Record Ids |
| ... | ... |
| Organico Agro | 22, 34, 43, 65, 76 |
| Moriko | 5, 230, 432, 678 |
| Rus Organics | 78, 475, 583, 598 |
| ... | ... |

| Global | |
|---|---|
| Manufacturers | Machine Ids |
| ... | ... |
| Organico Agro | 10 |
| VedShakti | 15, 16 |
| Sovam | 11, 12, 13 |
| ... | ... |

The upper table represents the situation when individual machines have their own index for the manufacturer field. Since it is present in every system that stores information for that table, this index is known as the local index. By searching for Organico Agro in the index, we could learn which items contained the data from this table. In spite of the fact that the primary query will hit all machines, the index is local, therefore the lookup will be lightning fast and the empty results will be back in no time. The second table takes a different approach by placing the index on the host computer. This index remembers which computer the record containing a certain value was initially stored in. Since we know Organico Agro is only on machine 10, we can save time and money by not checking out the other vending machines. Both indexing strategies are available for use at any time. Although queries may take up more space, this will speed them up.

### 3.2.2 JSON Data Query Using MongoDB and Aerospike

As discussed in Chapter 2, JSON is a type of semi-structured data. In the JSON structure, the atomic unit is a key-value pair. In order to query a

```
[
    {
        _id:1,                                          # Key-value pair

        name: "ram",

        age: 18,

        favorites: { artist: "Picasso", food: " porridge " },   # Named Tuple

        badges: [ "blue", "black" ],                    # Named Array

        points: [                                       # Named  Array  of  unnamed
                                                        Tuples
        { points: 85, bonus: 20 },

        { points: 75, bonus: 10 }

        ]

    }


    {

        _id:2,

        name: "bhusan",

        age: 21

    }

]
```

**Figure 3.1.**   JSON Data File

key-value pair, we need to be capable of carrying out one primary operation, which is to return the value after being provided with the key. This means that the value might potentially be an array or a list. Thus, query operations on it can either focus on its location on the list or its value. Both options are available. Let us take a sample JSON data file shown in Figure 3.1 to understand how to query JSON data [3].

As can be seen in Figure 3.1, the group of documents takes the form of a two-element array denoted by square brackets. As there is no key for the top-level array, it is prefixed with square brackets and given the default name db. The difference between a MongoDB query and a standard SQL query is that the latter specifies which sections of which documents inside a document collection should be returned, while the former specifies which parts of which records from one or more tables should be reported. As can be seen in Figure 3.2 [4], the most fundamental query in

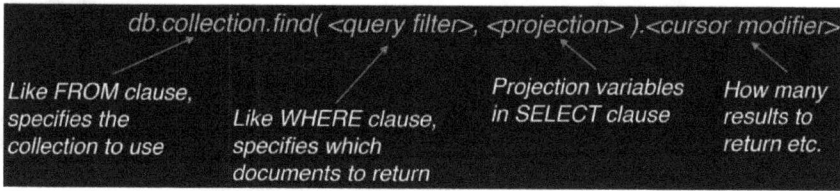

```
db.collection.find( <query filter>, <projection> ).<cursor modifier>
```

| | | | |
|---|---|---|---|
| Like FROM clause, specifies the collection to use | Like WHERE clause, specifies which documents to return | Projection variables in SELECT clause | How many results to return etc. |

**Figure 3.2.** MongoDB Query

MongoDB is a search function with two inputs and an optional quantifier.

There are certain key terms in the query shown in Figure 3.2. First, there is the keyword collection, which tells the system to make use of the particular document, and when limited to one table, it functions similar to the FROM clause. So, if the collection's name is drinks, db.drinks.find represents the first part. Query filter represents the second keyword, which functions like a WHERE clause, as it lists all the conditions the retrieved documents must satisfy, and in order to extract all of the results, then we need to leave this filter blank. The third keyword is known as a projection clause, and it consists of a set of variables that should be included in the result. The final keyword is called a cursor modifier which is separated by a dot from the find function. The term "cursor" has its roots in SQL, when it is referred to as a group of findings that are presented to the user in a single unit. If the table of outcomes is too long to be presented all at once, then it becomes important for the user to specify the amount of outcomes they would like to view or what proportion of the total results they desire.

To compare the SQL and MongoDB queries, let us look at some examples, as shown in Table 3.3. Let's say the first query wants everything from "drinks". The drink and price variables must be returned by the second query. The search feature in a MongoDB query requires an empty query condition, indicated by {}, although the projection clauses are named. For each attribute, 1 indicates its output and 0 indicates its absence. Only variables with a value of 1 are allowed by convention. Every document in MongoDB has an identifier called _id, and by default, each query outputs that object's id. In this case, we utilize the value 0. It should be supplied as _id:0 if we do not want it to return this attribute.

The third query has WHERE clause, which has a name that is equal to a value. In MongoDB, an equal to sign ('=') becomes a variable

**Table 3.3.**    Comparison of Queries in SQL and MongoDB

| | |
|---|---|
| **Query 1** | |
| *SQL* | SELECT * FROM Drinks |
| *MongoDB* | db.Drinks.find() |
| | |
| **Query 2** | |
| *SQL* | SELECT drink, price FROM Sells |
| *MongoDB* | db.Sells.find({},{drink:1, price:1}) |
| | |
| **Query 3** | |
| *SQL* | SELECT manufacturer FROM Drinks WHERE name='Organico Agro' |
| *MongoDB* | db.Drinks.find({name: "Organico Agro"},{manufacturer:1, id:0}) |
| | |
| **Query 4** | |
| *SQL* | SELECT DISTINCT drink, price FROM Sells WHERE price > 50 |
| *MongoDB* | db.Sells.distinct({price:{$gt: 50}},{drink:1, price:1,id:0}) |

delimiter (':'). A non-equality criterion is introduced in the fourth query, where the price must be more than 50. This query shows how operators may be used in queries in MongoDB. It has the variable followed by ":" which is then followed by the operator name, and the value being compared.

MongoDB provides a variety of features to extract the required data, such as searching using regular expressions, performing array slicing, and supporting compound statements [4]. In SQL, a compound expression is a query that combines many query conditions using an AND or OR operator. Now, a crucial aspect of semi-structured data is its ability to nest, which may be represented graphically as a tree structure. Now, the question is: What happens if we choose a tree node and ask for all of its descendants? In order to accomplish this, the system would have to recursively obtain child nodes while raising depth from the current node. However, recursive search is not supported by MongoDB.

From MongoDB, we will now go to Aerospike, which is key-value storage. Aerospike provides both programmatic access and a limited amount of query access to data. The data model for Aerospike is shown in Figure 3.3 [5].

Namespaces (ns) are the top-level data containers used to organize data that can reside in memory or on flash drives. Records, indexes, and

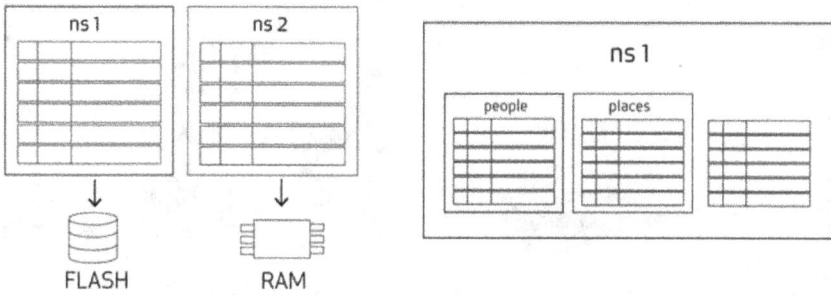

**Figure 3.3.** Aerospike Data Model [5]

policies are all contained in namespaces. Policies control namespace operations, such as data storage, including how much data are kept in RAM vs. on disc, how many clones of a record are kept, and when records expire. A namespace can contain sets, which can be considered as tables. Figure 3.3 shows two sets, people and places, and a set of records that are not in any set. Within a record, data are kept in one or more bins. Bins consist of a name and a value. Like SQL, Aerospike Query Language (AQL) is used to query the key-value data handled by Aerospike.

## 3.3 Big Data Integration

The problem of utilizing several diverse information sources to accomplish a task is known as data integration [6]. This section discusses the impact of increasing the number of data sources and the necessity of using data compression and defines record linkage, data interchange, and data fusion with the help of some use cases.

Due to the changing market dynamics, companies are constantly selling off a portion of their business or purchasing another business. In the commercial world of today, that is a relatively typical occurrence. Databases created and housed independently in various businesses would now need to be combined when these mergers and acquisitions occur. Let's make up a fictitious situation. Let's say a company can access two data sources from two distinct financial service providers, as shown in Figure 3.4.

As shown in Figure 3.4, the first data source is an insurance company that uses a relational DBMS to store its data. This database comprises nine tables, with a policy as the main informational item. The firm offers a

| Insurance Company's Partial Schema | Bank's Partial Schema |
|---|---|
| Policies(*PolicyKey*, *PolicyTypeKey*, Agent, Conditions)<br>PolicySales(*PolicyKey*, PolicyholderKey, StartDate,<br>    *TransactKey*,Premium,CoveragePeriod,<br>    CoverageLimit)<br>Transactions(*TransactKey*, Date, Time, Amount,<br>Balance)<br>Policyholders(*PolicyHolderKey*, Name, Address,<br>    City, State, ZIP)<br>Claims(*PolicyKey*, *ClaimKey*, *TransactKey*,<br>    ClaimAmount)<br>ClaimDescription(*ClaimKey*, *TypeKey*, *ClaimantKey*,<br>    ProcCode, Description)<br>Claimants(*ClaimantKey*, Name, Address, City, State,<br>ZIP)<br>ClaimTypes(*TypeKey*, Description)<br>PolicyTypes(*PolicyTypeKey*, Name, Description) | Accounts(*AcctNumber*, *AcctType*, *MemberID*,<br>    MemberType, TypeID, StartDate, EndDate,<br>    InterestRate, CreditLimit)<br>Individuals(*MemberID*, FName, MI, LName, SSN,<br>    Nationality, DoB, LegalStatus,<br>    FullAddress, Phone, PhoneType, Email)<br>Corporations(*MemberID*, Name, RegisteredAddress,<br>    CorporationType, Signatory1,<br>    Signatory2, DNBNumber, Phone, Email)<br>Transactions(*TrID*, AcctNum, Date, Time,<br>TransactionType,<br>    Description, TransactionAmount,<br>    Debit/Credit, Balance, Payoff)<br>AccountType(*TypeID*, Name, Description)<br>TransactionTypes(*Ttype*, Name, Description)<br>Disputes(*AccntNumber*, *DisputeID*, *TrID*, Date,<br>    DisputeAmt, Explanation, Valid, ValidatorID) |

**Figure 3.4.**   Data Sources from Two Hypothetical Companies

wide variety of insurance products marketed to individual customers via its agents. Policyholders pay their monthly dues, and claims are made as per their insurance policies. When they do, the database is kept up to date with the claims' specifics. The transaction is noted in the transactions table once the claims have been paid to the claimants. In this hypothetical scenario, the second company is a bank that also uses relational databases. Both individuals and corporations can open accounts at this bank. Now, accounts come in a variety of forms. A money market account, for instance, differs from a savings account. Additionally, a bank keeps track of all of its transactions in a table that might be rather extensive. However, a bank record dispute arises when a consumer charges the bank or declines to accept responsibility for the charge. This may occur, for instance, if a customer's online account was compromised or their debit card was stolen. In a disputes table, the bank does indeed keep a record of these irregularities and fraudulent activities. Let's see what happens after the data from these two companies are integrated. The company plans to undergo a promotional activity after the merger. If the owners of their insurance policies are also clients of the recently acquired bank, they would want to provide them some discount. Now, the question is how to identify such customers. An integrated view is created by querying two distinct data sources, which is represented by the new relation (discount-Candidates), as shown in Figure 3.5. It is integrated because the information is gathered from several sources, and it is referred to as a view because, in the context of databases, it is a relation that is computed from other relations.

Policyholders(*PolicyHolderKey*, Name, Address, City, State, ZIP)

Individuals(*MemberID*, FName, MI, LName, SSN, Nationality, DoB, LegalStatus, FullAddress, Phone, PhoneType, Email)

discountCandidates(*custID*, address, policyKey, AcctNumber)

PolicySales(*PolicyKey*, PolicyholderKey, StartDate, *TransactKey*, Premium, CoveragePeriod, CoverageLimit)

Accounts(*AcctNumber*, AcctType, *MemberID*, MemberType, TypeID, StartDate, EndDate, InterestRate, CreditLimit)

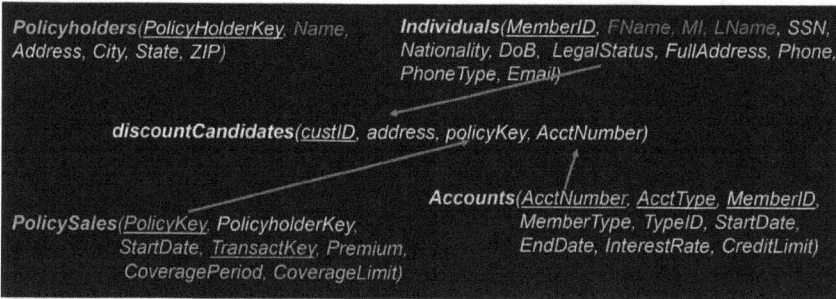

**Figure 3.5.** Integrated View by Joining Two Data Sources

Schema mapping is the process that is followed to populate the integrated view. The term "mapping" refers to the process of establishing correspondence between the attributes of the source relations and the attributes of the view, which is also referred to as a target relation. For instance, we may map each customer's FullAddress to the discountCandidates address field, but only if their names and addresses are the same in the two databases. However, there can be a problem called the record linkage while merging. To understand this problem, consider the records marked by the three arrows, as shown in Figure 3.6. The names and addresses of the two bank accounts and the policy document differ. Therefore, our earlier method would discard them. However, if we closely look at these records, we might consider it this way: Perhaps, this woman has changed addresses and has maiden and married names. This is called the record linkage problem. We want to ensure that the collection of data records that belong to an entity should be recognized in the integration process. This can be accomplished by clustering the values of different attributes or by using a set of matching rules.

As we've seen, the task of mapping the schema from two sources involves figuring out how its elements would connect to one another and how they would map the target schema. We also saw that producing one integrated connection out of a few relations from each source is not a straightforward operation.

In the big data scenario, there may be several hundred tables in each data source in a big data scenario, and there may even be more sources due to the company's expansion. Since there are many possible combinations, it becomes exceedingly challenging to solve this correspondence-making issue entirely and precisely.

**Figure 3.6.**   Record Linkage Problem in Data Integration

### 3.3.1 Big Data Integration Problems

This use case applies to companies that create customer-focused products. They want to know how their consumers react to the products, how successful their product marketing campaigns are, what kinds of issues customers are having, what new features or feature enhancements they desire, and so on. However, how do such organizations obtain this information and what kinds of data sources might do so? User surveys, emails from consumers, blogs and product review forums, specialist social media groups, and user forums are examples of the sources that might be used [7]. In a nutshell, these information sources may be found online or in content obtained from it, and their number may vary as new websites, articles, and discussion threads are constantly being created.

We need to look at the data fusion task first to cast the given problem of multichannel consumer analytics as a type of big data problem [8]. Let $S$ represent the set of data sources as mentioned above and $D$ represent the set of data items. A data item reflects a specific feature of a real-world object. Here, in this use case, a company's product, a part of a product, or a product feature refers to the data items. The source may or may not necessarily offer a value for each data item. Data fusion aims to find the values of data items from a source. Data fusion enables us to get the values of real-world objects from a subset of data sources because not all data sources will have relevant information about the data item. Now, one clear issue with the internet is that there are always too many data sources available, which causes several problems. First, it must be realized that there will be a lot of values for the same item if there are too many data sources. These will frequently be different, and occasionally, they will contradict. Using a voting system in this situation is a common strategy. However, issues with the data source might make even a voting system complicated. Trying to determine how reliable a source is is one of the challenges. We must determine if each data source accurately reports certain fundamental or well-known facts. If a data source mentions a rainbow-colored iPhone, it loses credibility because the value of the information it provides is false. As a result, a greater vote total can be given to a more reliable source. The second aspect is copy detection. Determining if one source has copied information from another might be crucial for data fusion in consumer analytics. The best outcomes for data integration may only be achieved if we have a method of evaluating the reliability of sources before information integration. Like in this use case, data come from too many redundant, possibly unreliable sources like the internet. Before performing information integration, there is a need to choose reliable and valuable sources out of all the available sources, and this problem is often called source selection.

The cost–benefit analysis of data fusion is shown in Figure 3.7, where the x-axis represents the total number of sources used and the y-axis

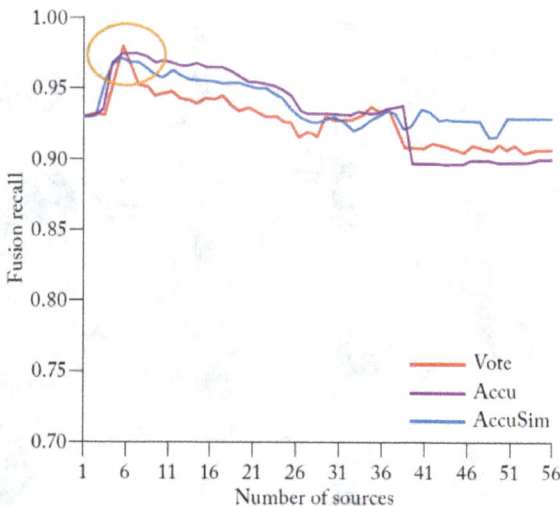

**Figure 3.7.**    Cost–Benefit Analysis of Data Fusion

calculates the percentage of true results returned [9]. It can be observed from Figure 3.7 that the efficiency decreases as the number of sources increases, with the plot peaking around six to eight sources. In a cost–benefit analysis, the cost includes both human and computational costs, while the benefit is a function of the accuracy of the fusion result.

## 3.4 Big Data Processing Pipeline

Most uses of big data involve a series of activities that are "pipelined" together. This conduit is used to transport data and undergoes several operations and transformations leading to the final product [10]. The term "pipe" was originally used to describe a UNIX action in which the results of one program were used as input for another. Let's consider a WordCounter in MapReduce [11]. It has one or more text files and counts how many times each word appears.

As shown in Figure 3.8, the files were initially divided into partitions of one file or multiple files, and stored in HDFS cluster nodes. Then each node underwent a Map operation, a user-defined function to count words. The map's generated key values were sorted alphabetically by key, and all the keys that shared a common term were grouped together into a single node. The values for key-value pairs with the same keys were then added

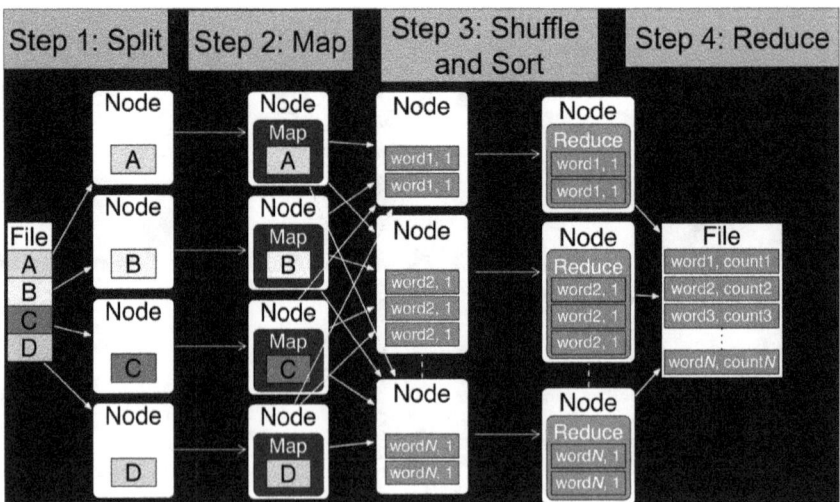

**Figure 3.8.** WordCounter in Hadoop MapReduce [11]

using the Reduce operation on these nodes. Figure 3.8 shows that there are in fact four separate processes, which are designated as follows: the data split step, the map step, the shuffle-sort step, and the reduction step. To achieve data parallel scalability, these four tasks can be generalized into three stages. In the first stage, we split the data. In the second stage, some operation is performed parallelly on the data, and finally, in the third stage, the results of the operations are combined to get the final product, like, in this example, the word count for each word. These three steps are often called split–do–merge. These three steps are stitched together for processing big data, thus making what we call a big data processing pipeline. When processing large amounts of data, the parallelism that occurs at each stage of the pipeline is mostly data parallelism. We can simplify the concept of data parallelism by defining it as the process of executing the same functions concurrently for the components or partitions of a dataset on different cores.

### 3.4.1 Data Transformation Operations in Big Data Processing Pipeline

Data are processed by a range of operations in data processing pipelines, which might apply a specific function to the data, convert the data to another format, combine the data with other datasets, or filter out some values from a dataset. The following section discusses the common data transformations used in big data pipelines.

#### 3.4.1.1 *Map and Reduce Operations*

One of the fundamental units of the big data pipeline is the Map operation. A Map operation is especially helpful when you want to apply the same operation or procedure to every member in a collection, such as a 10% hike in DA to each employee's salary. The Reduce operation then enables us to collectively apply the same process to objects of similar nature. The Reduce operation, for instance, comes in handy when we want to add our monthly spending across various categories, such as groceries, gas, and dining out. The graphical representation of the Map and Reduce operations is shown in Figure 3.9 [12]. The application of a Map function to data depicted in blue color is shown in Figure 3.9. Here, red, yellow, and green are keys for identifying each dataset. It is clear from the figure that

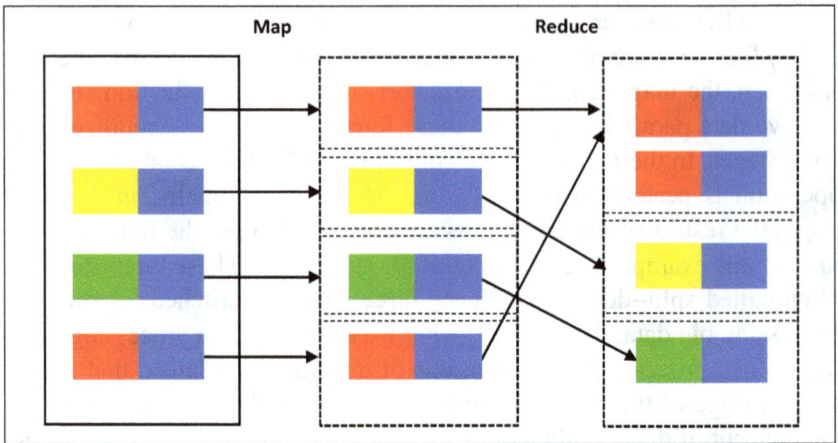

**Figure 3.9.**    Map and Reduce Operations

operations on each dataset are carried out in a distinct fashion, even for the keys that are the same color. In addition, we can see that the blue datasets that have the same color keys have been grouped by means of a Reduce function.

Both Map and Reduce are examples of different types of transformations that can operate on a single list of key and data pairings, as demonstrated on the left of Figure 3.9.

### 3.4.1.2 *Aggregation Operations*

Aggregation operation performs a specified transformation on a data collection by taking the relevant data pieces into consideration [13]. Let's say there is a bunch of circles of different colors. Here, the circle represents the data points and their color signifies the variety, as shown in Figure 3.10. If we apply a transformation, "$f$" that takes all the elements of data as input is called "aggregation". The most basic aggregation operation is the summation over all the data points. For example, in Figure 3.10, if we assume "$f$" to be a summation operation, it will output 17, which is the total number of data points.

Other variations of summation could return a sum by grouping data points of different types. For example, in Figure 3.10, it will output 3 (yellow), 3 (red), 4 (green), 4 (grey) and 3 (blue). This example can be related to a practical scenario where an organization is monitoring the

**Figure 3.10.**   Aggregation Operation

sales of different products. Here, one could relate the color to a different product type and these numbers (3,3,4,4,3), i.e., aggregation over colors, could be related to revenue generated by a product in each city when the product is sold. Another type of aggregation performs average over items of similar kind, which could be related to average revenue per product type. The other aggregation operations such as maximum, minimum, and standard deviation are simple yet quite helpful in drawing conclusions from big datasets. The aggregation operations can be performed as a sequence of operations, such as the maximum of the sums per product. That is, if we first add up sales for each product in each city and then apply the maximum function to the summation function's output, it will output the product that has the maximum sale in the whole country.

By selecting the appropriate aggregation operation, one may create concise and insightful insights that facilitate more rapid and efficient corporate decision-making. Aggregation operation typically produces smaller output datasets. Aggregation is thus a crucial tool when working with big data pipelines and massive amounts of data.

### 3.4.1.3 *Analytical Operations*

Analytical operations are used to transform the data into insights for making informed decisions [14]. Analytical operations are used to examine data in order to find significant trends and patterns that may be utilized to acquire knowledge about the problem under consideration. The information obtained from acquired knowledge eventually leads to a better decision based on the data [15,16]. Here, we discuss some of the common analytical operations, namely, classification, clustering, path analysis and connectivity analysis.

*Classification*: The goal of the classification problem is to achieve the categorical target from the input data. For example, we want to determine the sentiments of the public toward the current government. It can be viewed as a binary classification problem with the following two categories: like and dislike. Other examples of classification include, but are not limited to, predicting whether tumor cells are benign or malignant, sorting handwritten numbers from zero to nine, and validating or rejecting credit card transactions. A simple classification algorithm is the decision tree.

*Cluster Analysis*: The objective of clustering is to put together objects that are related in some way. For example, customers are categorized by their shared tastes in film subgenres. With the help of clustering, one can build good recommendation systems that can direct useful resources to the customers to spark boost sales. $k$-Means clustering is a popular and straightforward approach to data clustering, which divides the data into $k$ clusters, as shown in Figure 3.11 [17].

By employing similarity metrics like distance, this clustering aims to reduce the variation among the samples inside the same cluster. In Figure 3.11, the $k$ value is 3, i.e., it divides the data into three clusters.

Some of the other examples of clustering are stated as follows: For more effective targeted marketing, cluster analysis may also be used to divide a company's client base into discrete categories. It can also be used to discover articles or websites with comparable subjects in order to retrieve pertinent information [18]. For efficient administration of law enforcement resources, it is important to identify crime-prone locations in the city [19] and groupings of weather patterns, such as wet, chilly, or snowy.

*Connectivity Analysis*: Identifying and following groups in order to ascertain interconnections between things is what connectivity analysis is

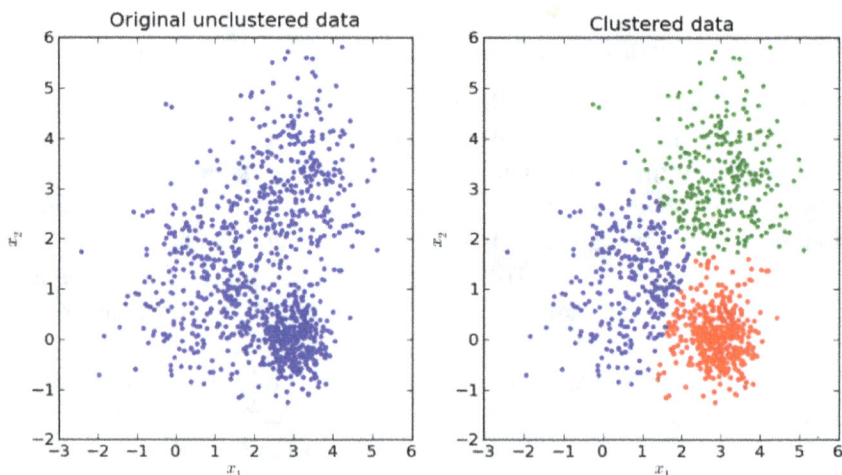

**Figure 3.11.**   $k$-Means Clustering

all about. Connections between entities are stronger in highly interactive groups than in less interactive groups. By analyzing the number of tweets and retweets, for instance, you can find out who is driving the debate about a certain topic and who the influencers are. Using this information, one can find the fewest number of influential individuals for social media marketing or political campaigns, for instance.

## 3.5 Big Data Management and Processing Using Splunk and Datameer

In the previous section, we discussed data integration, querying integrated data, and the issues in big data integration, and big data processing. Some of the data integration and processing tasks can be done using graphical user interface (GUI) based tools. In this section, we discuss two such tools: Splunk and Datameer.

### 3.5.1 Splunk

Whenever we think about the management of big data, we can think about Splunk. Machine data is the big data generated by applications, sensors, websites, and mobile devices. Machine data is omnipresent, and it is expanding at a fast rate. It is complex and unstructured but precious because it has a complete record of all behaviors and activities. By collecting and indexing this massive data regardless of the source or format, Splunk made its presence and helped unlock unprecedented value for thousands of customers worldwide. It supports users in monitoring, searching, analyzing, and reporting on their data quickly, conveniently, and in real time [20,21]. It is an application that turns raw data into insights. It derives values from big data sources. The architecture of the tool includes integrated data sources on which actions like collection, searching, queries, indexing, and visualization of data from the central location are performed. Splunk can be integrated with NoSQL and relational databases, and establishes a connection with the database administration workflow.

Several add-ons to the Splunk tool are also available. These are as follows:

- *Splunk Analytics for Hadoop*: Data stored in HDFS clusters can be accessed by setting up virtual indexing using the Splunk tool. This enables data processing, reporting, and visualization.

- *ODBC Driver*: This enables one to connect Splunk enterprise edition and tools, such as Excel and Tableau.
- *Database Connect*: This add-on integrates database applications, such as MySQL, Oracle, SAP SQL, and DB2, with Splunk.

There are several benefits of using the Splunk tool to manage big data:

- It enables the user to process huge data (i.e., hundreds of terabytes of daily data) collected from different sources like data centers and the cloud.
- It provides insight across real-time data and historical data.
- It adds context by extending it to relational data. This enables the user to integrate data from data warehouses and relational databases.

### 3.5.2 Datameer

Datameer has been developed on the open-source Hadoop project. It is considered as a big data analytics platform for Hadoop-as-a-service. As a tool, it has established itself as an *end-to-end analytics platform*. It uses Hadoop as the base to simplify the big data analytics environment. It works in the Hadoop environment and prepares data ingested from HDFS, and then queries are applied to the data using Hive or Spark. Finally, visualization of the queried data can be done in order to under-stand the pattern of information. In view of SaaS, Datameer targets department-level deployments [22]. It includes more than 70 data con-nectors to handle different types of data and their sources and more than 270 pre-built points and click analytics functions. It provides quick insights by empowering direct control of the data via the use of data con-nectors and performing different analytics targeting customers, opera-tions, and data security.

## 3.6 Chapter Summary

Big data processing is the process of handling large and complex datasets that cannot be processed using traditional data processing techniques. It involves capturing, storing, processing, analyzing, and visualizing large volumes of data to extract valuable insights and information. One of the primary challenges of big data processing is the volume of data, which can be in petabytes or even exabytes. To handle such large volumes of data, new technologies such as Hadoop, Spark, and NoSQL databases have emerged. These technologies allow for distributed processing, parallel computing, and fault tolerance, enabling businesses to process and analyze massive datasets efficiently. The benefits of big data processing are numerous. It enables businesses to make data-driven decisions, identify new opportunities, and optimize their operations. It also allows for better customer insights, improved product development, and enhanced cybersecurity.

In conclusion, big data processing is a crucial aspect of modern business operations. With the explosion of data, businesses need to develop the necessary tools and techniques to manage and extract insights from large and complex datasets. The ability to process and analyze big data is essential for companies to remain competitive and thrive in the digital age.

# References

1. Prasanth, T. & Gunasekaran, M. (2019). Effective big data retrieval using deep learning modified neural networks. *Mobile Networks and Applications*, 24, 282–294.

2. Ilyas, I. F., Beskales, G., & Soliman, M. A. (2008). A survey of top-k query processing techniques in relational database systems. *ACM Computing Surveys (CSUR)*, 40(4), 1–58.

3. Yusof, M. K. & Man, M. (2017). Efficiency of JSON for data retrieval in big data. *Indonesian Journal of Electrical Engineering and Computer Science*, 7(1), 250–262.

4. Botoeva, E., Calvanese, D., Cogrel, B., & Xiao, G. (2018). Expressivity and complexity of MongoDB queries. In *21st International Conference on Database Theory (ICDT 2018)*, 26–29 March 2018, Vienna, Austria. Schloss Dagstuhl-Leibniz-Zentrum fuer Informatik.

5. Srinivasan, V., Bulkowski B., Chu W. L., Sayyaparaju S., Gooding A., Iyer R., Shinde A., and Lopatic T. (2016). Aerospike: Architecture of a real-time operational dbms. *Proceedings of the VLDB Endowment*, 9(13), 1389–1400.

6. Dong, X. L. & Srivastava, D. (2013). Big data integration. In *2013 IEEE 29th International Conference on Data Engineering (ICDE)* (pp. 1245–1248), 8–12 April 2013, Brisbane, QLD, Australia. IEEE.

7. Kadadi, A., Agrawal, R., Nyamful, C., & Atiq, R. (2014, October). Challenges of data integration and interoperability in big data. In *2014 IEEE International Conference on Big Data (Big Data)* (pp. 38–40), 27–30 October 2014, Washington DC, USA. IEEE.

8. Khan, A. & Colace, F. (2021). "Knowledge Graph: Applications with ML and AI and Open-Source Database Links in 2022", Insights2Techinfo, p. 1. https://insights2techinfo.com/knowledge-graph-applications-with-ml-and-ai-and-open-source-database-links-in-2022/.

9. Shahandashti, S. M., Razavi, S. N., Soibelman, L., Berges, M., Caldas, C. H., Brilakis, I., ... & Zhu, Z. (2011). Data-fusion approaches and applications for construction engineering. *Journal of Construction Engineering and Management*, 137(10), 863–869.

10. Javed, M. H., Lu, X., & Panda, D. K. (2017, December). Characterization of big data stream processing pipeline: A case study using Flink and Kafka. In *Proceedings of the Fourth IEEE/ACM International Conference on Big Data Computing, Applications and Technologies* (pp. 1–10), 5–8 December 2017, Austin, Texas, USA.

11. Issa, J. A. (2015). Performance evaluation and estimation model using regression method for Hadoop WordCount. *IEEE Access*, 3, 2784–2793.

12. Dittrich, J. & Quiané-Ruiz, J. A. (2012). Efficient big data processing in Hadoop MapReduce. *Proceedings of the VLDB Endowment*, 5(12), 2014–2015.

13. Gama, J. & Rodrigues, P. P. (2007). Data stream processing. In *Learning from Data Streams: Processing Techniques in Sensor Networks* (pp. 25–39), 2007, Springer, Berlin, Heidelberg, New York.

14. Kibria, M. G., Nguyen, K., Villardi, G. P., Zhao, O., Ishizu, K., & Kojima, F. (2018). Big data analytics, machine learning, and artificial intelligence in next-generation wireless networks. *IEEE Access*, 6, 32328–32338.

15. Din, S., Paul, A., Ahmad, A., Gupta, B. B., & Rho, S. (2018). Service orchestration of optimizing continuous features in industrial surveillance using big data based fog-enabled internet of things. *IEEE Access*, 6, 21582–21591.

16. Casillo, M., Gupta, B. B., Lombardi, M., Lorusso, A., Santaniello, D., & Valentino, C. (2022). Context aware recommender systems: A novel approach based on matrix factorization and contextual bias. *Electronics*, 11(7), 1003.

17. Duran, B. S. & Odell, P. L. (2013). *Cluster Analysis: A Survey* (Vol. 100). Springer Science & Business Media. Springer, Berlin, Heidelberg, New York.

18. Alsmirat, M. A., Jararweh, Y., Al-Ayyoub, M., Shehab, M. A., & Gupta, B. B. (2017). Accelerating compute intensive medical imaging segmentation algorithms using hybrid CPU-GPU implementations. *Multimedia Tools and Applications*, 76, 3537–3555.

19. Plageras, A. P., Psannis, K. E., Stergiou, C., Wang, H., & Gupta, B. B. (2018). Efficient IoT-based sensor BIG Data collection–processing and analysis in smart buildings. *Future Generation Computer Systems*, 82, 349–357.

20. Zadrozny, P. & Kodali, R. (2013). *Big Data Analytics Using Splunk: Deriving Operational Intelligence from Social Media, Machine Data, Existing Data Warehouses, and Other Real-time Streaming Sources.* Apress.
21. Zadrozny, P., Kodali, R., Zadrozny, P., & Kodali, R. (2013). Big data and Splunk. In *Big Data Analytics Using Splunk* (pp. 1–7).
22. Di Martino, B., Aversa, R., Cretella, G., Esposito, A., & Kołodziej, J. (2014). Big data (lost) in the cloud. *International Journal of Big Data Intelligence,* 1(1–2), 3–17.

# Chapter 4

# Big Data Analytics and Machine Learning

Big data's role in the success of many significant technological businesses is no secret. However, as more companies use it to store, analyze, and extract value from their massive volume of data, it becomes increasingly difficult for them to use the gathered information most effectively [1]. This is where machine learning proves to be helpful. Data always overpower the machine learning algorithm. No matter how good the learning algorithm is, it will be useless without a significant amount of relevant data. The efficiency and accuracy of these machine learning algorithms increase as we feed them with more relevant data. Data act as a blessing for machine learning algorithms [2]. Machine learning systems perform better as they get more and more data. Therefore, adopting machine learning for big data analytics is a natural step for businesses to optimize the adoption of big data. This chapter focuses on applications of machine learning to big data.

## 4.1 Introduction to Machine Learning

The study of how computers can acquire knowledge through observation and experience is known as "machine learning" [3]. The system or model acquires the ability to carry out a given task by studying several instances of that particular task in action. With numerous examples, a machine learning model can learn to identify a dog in a picture. The machine learning model can learn to do specific tasks itself without the need for step-by-step instruction or explicit programming. The model is trained to recognize canine-specific characteristics that help determine whether an image contains a dog or not. The model will have to be trained on the data to accomplish this; therefore, the quality and amount of data should be good so that the model can perform the task well. A model's ability to learn a task depends heavily on the quantity and quality of data used to construct it. Due to the fact that machine learning algorithms are data-driven, they can be utilized to reveal previously unseen connections, leading to valuable insights into the data.

Machine learning is an interdisciplinary field requiring the knowledge of mathematics and statistics, computer science and artificial intelligence, and the domain knowledge where one wants to apply the solution [4]. Understanding the domain, the data specific to the application, and the intended use of the results all constitute domain knowledge. Due to the importance of these factors in developing a successful machine learning model, domain expertise must be a fundamental part of any successful machine learning implementation.

Machine learning applications are widely used by us, either knowingly or unknowingly. For example, every time we make a purchase using a credit card, the transaction is compared to the previous credit card transactions to see if it is valid or perhaps fraudulent [5]. The transaction will be marked as suspicious if it differs significantly from the previous purchases. For example, the product being purchased falls within a specific class, in which one never expressed inclination or the point of sale will be located outside the country. In such events, we immediately get notifications to alert and get confirmation from us. Machine learning is widely used in situations like these. Machine learning is also used every day when a person uses an ATM and deposits a check made by hand. The machine learning algorithm reads the numbers from a check and uses that information to make a financial transaction. Recommendation at the famous shopping platforms is another application of machine learning that everyone has often encountered [6,7]. After purchasing on a website, we frequently receive a list of related products with captions, such as "as you like this item, you may also like these as well" or "customers who purchased this item also bought these things". A machine learning algorithm connects similar products to the item that was just purchased, and these similar products are displayed to the customer because the machine learning algorithm has determined that you might also be interested in the related items. This typical usage of machine learning is frequently seen in sales and marketing. Apart from these, machine learning has been applied in other areas, such as research, medicine, retail, law enforcement, education, and many more areas.

## 4.1.1 Machine Learning Techniques

There are several categories of machine learning approaches for various kinds of problems [8]. The primary categories are classification [9], regression [10], cluster analysis [11], and association analysis [12]. The objective of classification is to predict the category of incoming data. Predictions of sunshine, rain, wind, or clouds serve to highlight this point. Sensor readings for environmental variables like temperature, humidity, pressure, wind speed and direction, etc. would form the basis of the input data in this scenario. Other examples of classification include the determination of a benign or malignant tumor from medical imaging, forecasting if it will rain on the following day, determining if a loan application has a high, medium, or low risk, and identifying the favorable, negative, or neutral nature of a tweet or review. If the model requires determining a numeric value rather than a category, the problem becomes a regression problem. Other regression applications include estimating a product's demand depending on the time or season of the year, predicting a score on a test, assessing the possibility that treatment would be beneficial for a particular patient, and forecasting the amount of precipitation for an area [10].

Cluster analysis, also known as clustering, seeks to group data points that are similar [11]. Customer segmentation is one of the most frequent applications of cluster analysis. This indicates that you are segmenting your consumer base based on customer type. For instance, dividing your clients into elderly citizens, adults, and adolescents would be highly advantageous. These groups have distinct likes and dislikes and distinct purchase patterns. By dividing your clients into several groups, you may give more relevant advertisements to each group's specific interests. Other cluster analysis applications include identifying places with similar geographies, such as deserts, grasslands, and mountains, categorizing various kinds of tissues based on medical image analysis, and identifying distinct categories of weather patterns, including snowy, dry, and monsoon seasons.

Association analysis aims to develop a set of rules for capturing relationships between objects or occurrences [12]. The rules are used to identify when the particular objects co-occur. The term "market basket analysis" refers to the typical use of association analysis, which is utilized to comprehend customers' purchasing behavior. Association analysis can demonstrate, for instance, that bank clients who hold certificates of deposit are also interested in alternative investment vehicles, such as money market accounts. This data can be utilized for cross-selling. If you offer money market accounts to your CD-holding consumers, they will likely open one. A supermarket chain utilized association analysis to establish a link between two seemingly unrelated goods. They noticed that many clients who purchase diapers late on Sunday also purchase alcohol. This data was then utilized to put beer and diapers together, increasing sales for both products. This is the well-known link between diapers and beer. Other applications of association analysis include identifying items that are frequently purchased together, recommending similar items based on customers' purchasing behavior or browsing histories, and identifying web pages that are frequently accessed together so that you can provide these related web pages concurrently in a more efficient manner.

There is another categorization of machine learning techniques called supervised vs. unsupervised machine learning [13]. In supervised learning, the objective, or what the model attempts to predict, is given to the model. If we refer back to the example of forecasting a weather category to be bright, windy, rainy, or overcast, each sample in the dataset is assigned to one of these four categories. Therefore, the data are annotated, and forecasting the weather categories is a supervised task. Classification and regression are typically supervised methods.

In contrast, in unsupervised techniques, the goal being predicted by the model is unknown or unavailable. This indicates that your data are unlabeled. If we refer back to the example of segmenting clients into several groups using cluster analysis, the samples in the data are not tagged with the correct category. Instead, segmentation is accomplished using a clustering algorithm to group things according to their common properties. Thus, the data are unlabeled, and classifying clients into various categories is unsupervised. Cluster analysis and association analysis are often unsupervised methods.

## 4.2 Machine Learning Process

The machine learning process comprises the following stages, as shown in Figure 4.1. The process of machine learning is highly iterative [14].

**Figure 4.1.** Machine Learning Process

Findings from one phase may necessitate repeating a previous step with new information. During the prepare phase, if some data quality concerns are discovered, then there is a need to return to the acquire phase to resolve problems with data gathering or to collect extra data we missed the first time. Additionally, each process may need many iterations.

The initial stage in this process involves collecting all relevant data for the given task. Here, we must identify all data sources, gather the data, and then combine data from these multiple sources. The next stage in the process of machine learning is data preparation. This stage is further split into two parts: data exploration and data pre-processing. The initial data preparation phase entails a preliminary examination of the data to determine the type of data that must be utilized. The key focus while determining the nature of data is on its characteristics, format, and quality because a thorough understanding of data leads to a more accurate analysis and a more fruitful conclusion. After learning more about the data through exploratory research, the data must be prepared for analysis. This involves data cleansing, variable selection, and data transformation to make the data more appropriate for analysis in the subsequent stage. The produced data would subsequently be forwarded to the analysis phase. This process comprises selecting the appropriate analytical methodologies, developing a data-driven model, and evaluating the outcomes. Step four of machine learning involves communicating outcomes. This involves analyzing the findings in relation to the project's intended outcome. The final stage involves applying the results. The act step focuses mostly on determining actions based on analysis-derived insights. The detailed discussion for each stage will be covered in the subsequent sections.

Now, we discuss some basic terminologies which are frequently used in the context of machine learning. Consider Table 4.1, containing weather data, to understand these basic terminologies.

*Sample*: A data sample is an illustration of a data entity. This is a data row or a single record in a table. Each row in Table 4.1 is a sample of

**Table 4.1.**   Weather Data

| | | Variables | | | |
| --- | --- | --- | --- | --- | --- |
| Record ID | Date | MinTemp | MaxTemp | Rain fall | |
| 1 | 13/12/2018 | 8 | 23 | 7 mm | |
| 2 | 13/12/2019 | 7 | 28 | 6 mm | Samples |
| 3 | 13/12/2020 | 8 | 29 | 8 mm | |
| 4 | 13/12/2021 | 6 | 24 | 9 mm | |

weather data on a specific day. In the machine learning context, the sample is also called a record, example, row, instance, or observation.

*Variable*: Each sample in Table 4.1 is connected with five values. The sample ID, sample date, minimum temperature, maximum temperature, and daily precipitation are all values that are specific to that sample. These distinct values are referred to as sample variables. In the machine learning context, the variable term is often used interchangeably with a feature, column, dimension, attribute, or field. Each sample in the dataset is defined by these terms, which describe its unique properties. Each variable has an associated data type with it. Common types include numerical and category data.

Numeric variables, as their name indicates, are variables with numerical values. Measurement and classification of numerical variables are possible. It's important to keep in mind that a numerical variable can only store integers or continuous numbers. It could be entirely positive, entirely negative, or something in between. Some quantities are positive and continuous, like a person's height; others are positive integers, like the score on a test; yet others, like the number of transactions per hour, can be either positive or negative, depending on the context.

In contrast to numeric variables, categorical variables have labels, titles, or categories for values. Blue, black, yellow, white, etc. are all possible values for a variable that describes the color of an item, such as a car. These are values that can't be represented numerically but nonetheless describe some aspect of an object. The terms "qualitative" and "nominal" are also used to describe the category variables. Categorical variables can include things like a person's age, marital status, or the age ranges they fall into when shopping. Electronics, home improvement, and home decor are all examples of different types of products.

## 4.2.1 Acquire

The goal of this stage is to locate and compile all information that is relevant to the problem at hand. The first step is to catalog all of the essential information and where we can find it, as the data might originate from several sources, including files, databases, the internet, and mobile devices. Following the identification of the data and data sources, gathering and integrating the data from various sources is necessary. Different types of data formats may necessitate conversion. In addition, it may be necessary to align the data, as data from various sources may have varying temporal or spatial resolutions.

## 4.2.2 Prepare

After the collection of data, the subsequent stage involves the preparation of data for analysis. This stage consists of two components: data exploration and data pre-processing. The goal of data exploration is to develop an understanding of the nature of the data, and the goal of data pre-processing is to create data for analysis. Exploratory data analysis targets finding patterns, trends, correlations, and outliers by visualizing the data and finding the data's summary statistics. In pre-processing, three key tasks are performed: data cleaning, feature selection, and feature transformation.

### 4.2.2.1 *Exploratory Data Analysis (EDA)*

Exploration of data entails conducting a preliminary investigation to learn more about its unique qualities [15,16]. One possible goal of exploring data is to look for connections, trends, outliers, etc. The use of correlations allows one to see how various data points are connected to one another. The trends identify if the variables in the data follow an upward/increasing or downward/decreasing direction. There may be problems with the data or an interesting data point that demands more examination if there are any outliers. It's possible that we won't get the most out of the data until we take the time to explore it properly. Summary statistics provide numerical descriptions of the data and can be calculated for further exploration. Mean, median, mode, are the primary measures of central tendency while range, standard deviation are the primary measures of dispersion.

The most common number in a set of data is called the mode, whereas the range and the standard deviation quantify the spread of the data. These metrics will provide insight into the nature of your data. These can highlight the potential problem with the data. For instance, if the range of ages in the data contains negative digits or a figure much more than 100, something questionable in the data must be investigated. Further, visualization tools like histograms, line plots, heat maps, and scatter plots provide rapid and efficient data exploration. A histogram illustrating the data distribution might reveal skewness or unusual dispersion in outliers. A line plot can be utilized to examine data patterns. A heat map can indicate where the most popular locations are. A scatter diagram efficiently illustrates the relationship between two variables. Other such kinds of tools are quite helpful in developing a good understanding of the data.

Further details about the two key categories of the techniques used for performing exploratory data analysis are given in the subsequent sections.

### 4.2.2.1.1 Summary Statistics

Data values can be summarized with the help of summary statistics, which can reveal useful insights. Mean, median, and standard deviation are the three summary statistics used most frequently [17]. One way to summarize the information in a dataset is with a summary statistic, which is a single number. For instance, a mean is a single number that represents an average over any size collection. The mean, for example, can be interpreted as a measure of the geographic center of the dataset; as such, summary statistics provide a quick and easy way to get an overview of the entire dataset. Measures of location or centrality, measures of spread, measures of shape, and measures of dependency are the focus of this section.

*Measures of Location*: Location measures are descriptive statistics that pinpoint the heart of a dataset. Mean, median, and mode are all examples of statistical measures. The average or mean of the data collection is calculated by arithmetic mean. The median is the middle number regardless of the order in which the dataset is sorted. In a sorted set of numbers, half will be lower than the median and half will be higher than the median. The median is calculated by averaging the middle values when there are an even number of observations. The mode represents the most typical occurrence of a given value.

*Measures of Spread*: Dispersion metrics characterize a dataset's degree of diversity. Standard metrics of dispersion include the minimum, maximum, range, standard deviation, and variance. The smallest and greatest numbers in a set are called the minimum and maximum, respectively. A measure of data dispersion, range, is the difference between the data's extremes (highest and lowest values). The variability in a data collection can be measured by calculating its standard deviation. If the standard deviation of a dataset is small, then the samples are concentrated close to the mean. The data samples are likely to be scattered if the standard deviation is high. There is a close link between variance and standard deviation. As the square root of the standard deviation, variance measures how far individual data points are from the mean.

*Measures of Shape*: The form of the distribution of a set of values can be described using various measures of shape. Skewness and kurtosis are typical members of shape. Skewness determines whether data values follow asymmetric distribution, as shown in Figure 4.2.

As shown in Figure 4.2, a skewness score close to zero suggests that the data distribution is approximately normal. When the skewness score is negative, it indicates that the distribution is left-skewed. Conversely, if the skewness score is positive, it means that the data is right-skewed. The kurtosis statistic measures how heavy or fat the tails of a distribution of data are. Outliers are present when the kurtosis value is high, which describes a distribution with fatter tails and a higher peak in the middle. A smaller kurtosis number indicates that the distribution is more centered and has fewer outliers, whereas a bigger kurtosis value indicates that the tails are longer and thinner.

*Measures of Dependence*: Measures of dependency determine whether or not variables are related. The pairwise correlation is a standard measure of dependency. The correlation only applies to numerical variables, ranging from zero to one, with zero representing no connection and one representing a one-to-one relationship. For example, let the correlation of 0.89 come for the weight and height data. It gives a sense of a high correlation between weight and height, and the same is our expectation as

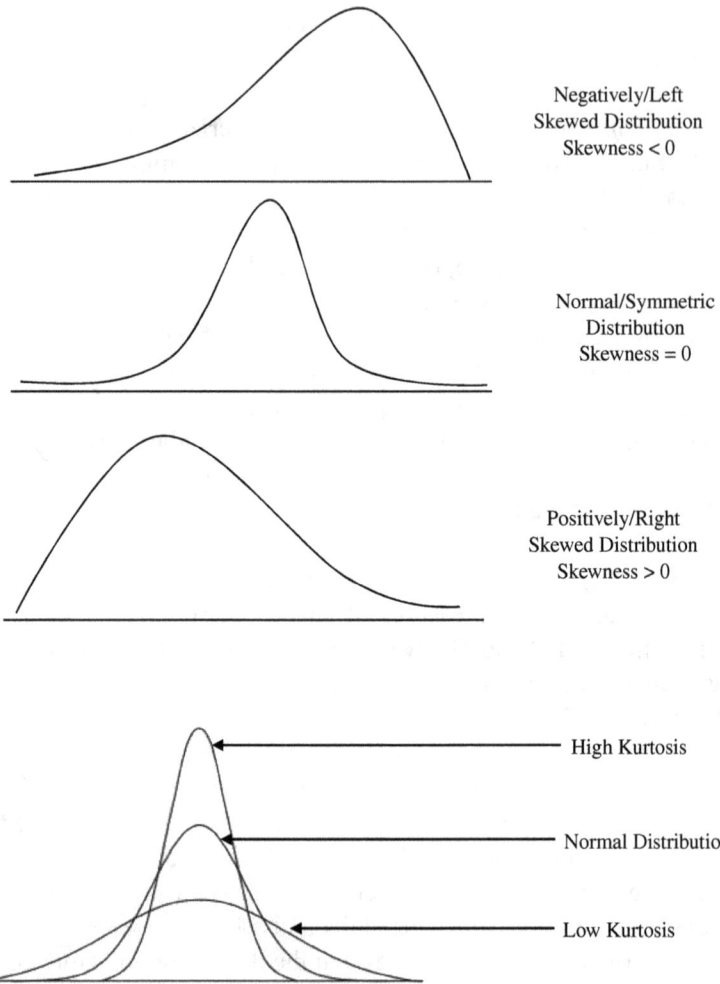

**Figure 4.2.**   Measures of Shapes

well; height and weight should go hand in hand because of the inherent relationship between them.

The statistical summaries that have been discussed up to this point apply to numerical variables. For categorical variables, as a measure of summary statistics, we want to identify the number of categories as well as the frequency with which each category occurs. A contingency table will allow us to accomplish this goal. Consider an example of categorical data, as shown in Table 4.2. It contains the distribution of pets and their colors. It is clear from the table that dogs make up the largest percentage of households with pets, while fish make up the smallest percentage.

Similarly, if we sort the pets by color, we find that black is the most common and orange is the least. The distribution between classes is also displayed in the contingency table. Only fish are orange, yet dogs make up the vast majority of brown pets.

With machine learning, we want to validate the data as rapidly as possible; therefore, in addition to checking the standard summary statistics for numerical variables and the category count for categorical variables, we also want to check particular supplementary statistics. One of the first things that must be done is a detailed check of the dataset to ensure that it contains the appropriate number of rows and columns. Does the row count correspond to the sample count that was anticipated? Does the column count correspond to the variable count that was anticipated?

Examining the values of the dataset's first and final few samples to determine if they are appropriate is another straightforward method for validating data. For example, consider the weather report shown in Table 4.3. In this, some preliminary checks could be as follows: are the data types of the variables correct, i.e., if the date is stored as a date or

**Table 4.2.**   Categorical Data

| Color/Pet | White | Brown | Black | Orange | Total |
|-----------|-------|-------|-------|--------|-------|
| Dog       | 34    | 44    | 32    | 0      | 110   |
| Cat       | 25    | 2     | 43    | 0      | 70    |
| Fish      | 1     | 0     | 5     | 33     | 39    |
| Total     | 60    | 46    | 80    | 33     | 219   |

**Table 4.3.**   Weather Report

| ID | Date       | MinTemp | MaxTemp | Rainfall |
|----|------------|---------|---------|----------|
| 1  | 2010-06-17 | 56      | 75      | 0.1      |
| 2  | 2016-06-18 | 52      | 78      | 0.0      |
| 3  | 2010-06-19 | 50      | 78      | 0.0      |
| 4  | 2010-06-20 | 54      | 77      | 0.0      |

a timestamp or if it is stored as a string or a numeric value? Which meas-urement system is being used to express the temperature: degrees Celsius or Fahrenheit? Are all of the figures for rainfall consistent with one another, or are there certain values that don't seem to belong?

Checking for missing data is another critical step. One must determine how many samples have missing values and the proportion of missing data. In the following sections, we discuss how to deal with missing values, which is a key stage in the data preparation process.

### 4.2.2.1.2  Visualization Methods

Visualizing data, i.e., examining it graphically, is a very effective method for analyzing data. Several types of plots may be used to visualize data, and among them, histograms, line plots, scatter plots, bar plots, and box plots are the most popular ones [17].

*Histogram*: The distribution of a variable can be displayed using some-thing called a histogram. The range of potential values for the variable is segmented into the number of bins, and then the count of the values that fit into each bin is performed; this counts the total amount of space occu-pied by the values in each bin, as shown in Figure 4.3.

A histogram may reveal a variety of information about a variable in the data, such as the variable's central tendency and the most common value for that variable. We can also determine the outliers from the

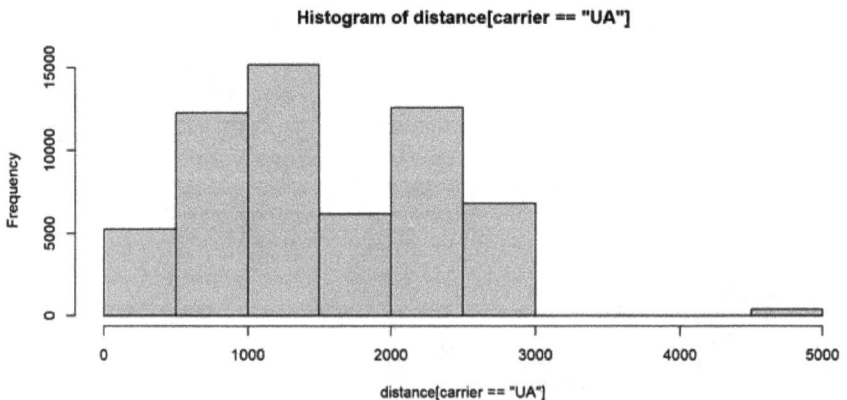

**Figure 4.3.**   Histogram Plot of Distance Covered by United Airlines

histogram; as shown in Figure 4.3, the bin at 5000 is an outlier. A histogram will also reveal if the values of the variable are skewed and in which direction the skewness is directed: either to the left, in the direction of smaller values, or to the right, in the direction of larger values, as shown in Figure 4.4.

*Line Plot*: A line or time-series plot illustrates the change of data values over time. The *y*-axis displays the values of a variable or variables, while the *x*-axis depicts the progression of time. The values of the data are plotted on the resultant line to show their progression through time. A line plot is a useful tool for illustrating trends among variables. For instance, a cyclical pattern may occur when the values begin high, decline, and rise again, as shown in Figure 4.5.

As seen in Figure 4.6, trends can also be identified in which the numbers vary but exhibit an overall increasing tendency over time.

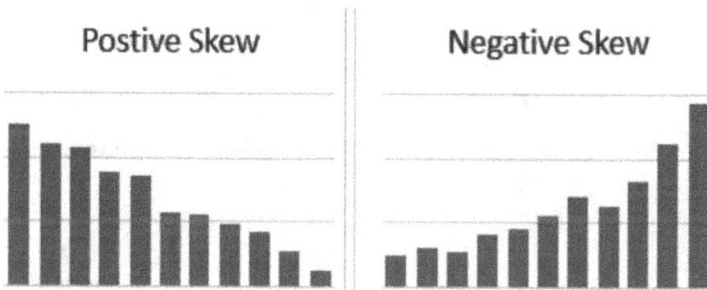

**Figure 4.4.** Determination of Skewness in Histogram

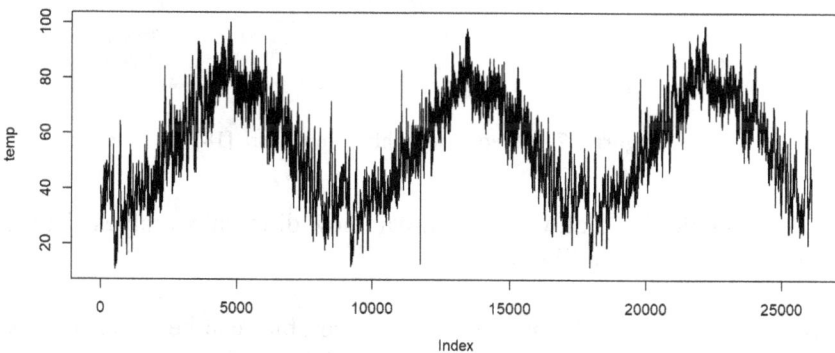

**Figure 4.5.** Line Plot

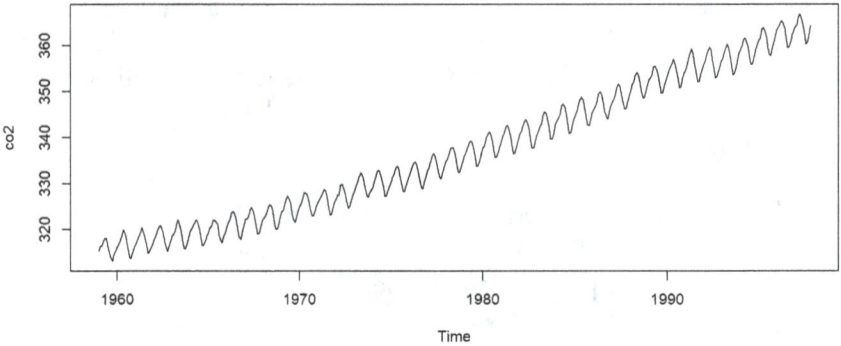

**Figure 4.6.**   Indication of Trends in Line Plots

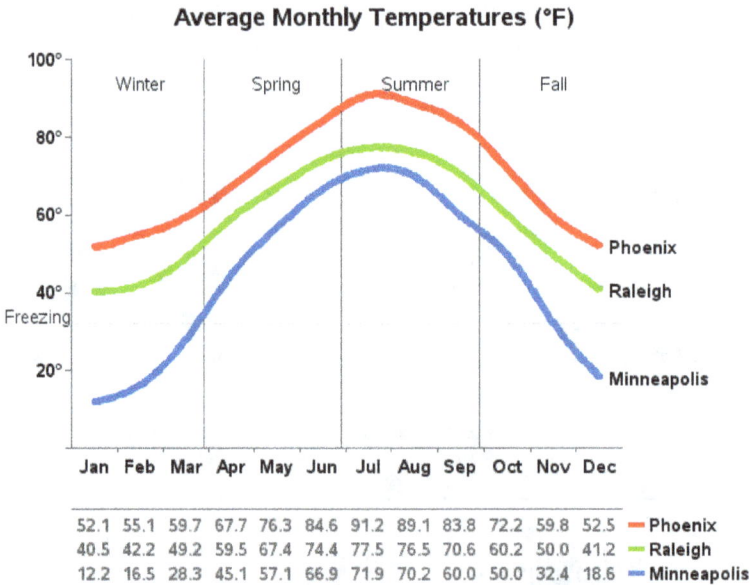

**Average Monthly Temperatures (°F)**

| | Jan | Feb | Mar | Apr | May | Jun | Jul | Aug | Sep | Oct | Nov | Dec | |
|---|---|---|---|---|---|---|---|---|---|---|---|---|---|
| | 52.1 | 55.1 | 59.7 | 67.7 | 76.3 | 84.6 | 91.2 | 89.1 | 83.8 | 72.2 | 59.8 | 52.5 | Phoenix |
| | 40.5 | 42.2 | 49.2 | 59.5 | 67.4 | 74.4 | 77.5 | 76.5 | 70.6 | 60.2 | 50.0 | 41.2 | Raleigh |
| | 12.2 | 16.5 | 28.3 | 45.1 | 57.1 | 66.9 | 71.9 | 70.2 | 60.0 | 50.0 | 32.4 | 18.6 | Minneapolis |

**Figure 4.7.**   Average Monthly Temperature [18]

On a single-line plot, we can compare how different variables change over time, as shown in Figure 4.7.

*Scatter Plot*: The correlation between two variables can be depicted very clearly through the use of a scatter plot, which is an effective way. On this graph, one variable is displayed along the *x*-axis, and the other variable is

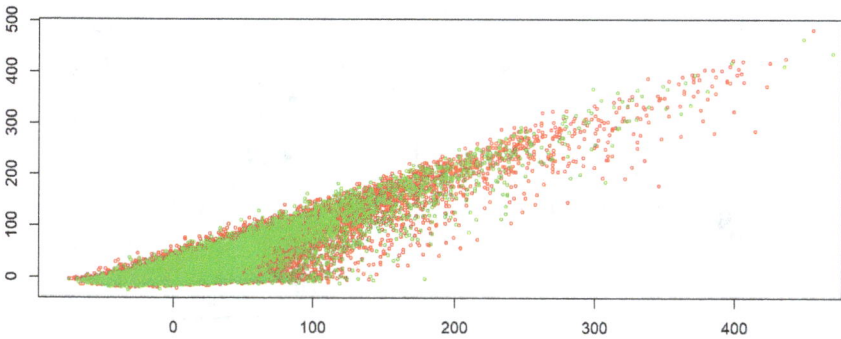

**Figure 4.8.** Scatter Plot of Arrival vs. Departure Time of United Airlines

displayed along the *y*-axis. The values of the two variables, *X* and *Y*, are plotted using each individual sample. The association between the two variables can be seen graphically in the scatter plot that was generated. For example, two variables, such as the arrival time and departure time of a flight, might have a positive correlation, as shown in Figure 4.8.

The correlation between two variables can be positive or negative, as shown in Figure 4.9. A positive correlation shows that when the value of one variable increases, the value of the other variable also increases at the same rate. The scatter plot in the upper right corner of Figure 4.9 displays a negative correlation between the two variables. This indicates that when one variable's value increases, the other variable's value decreases proportionally.

It is also possible for two variables to have a connection that is not linear, as shown in the bottom-left corner of Figure 4.9. This suggests that there is no guarantee that a change in one variable will correlate to the same change in the other variable. This is seen by the scatter plot, which consists of a curve rather than a straight line, which would be expected for a linear correlation. In addition, there is a possibility that there is no connection between the two variables, as seen in the lower right corner of Figure 4.9. In this situation, the dots in the scatter plot will be randomly distributed, showing no correlation between how the two variables vary in relation to one another.

*Bar Plot*: The distribution of categorical variables is illustrated using a bar plot. Another tool for analyzing the variable's value distribution is the

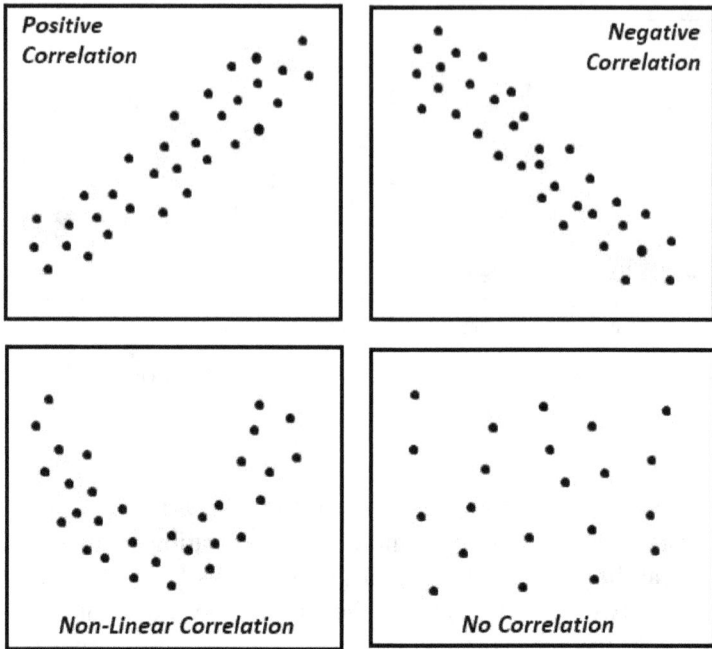

**Figure 4.9.**    Correlation Between Two Variables

histogram. A bar plot is more common for categorical variables, while a histogram is more common for numerical data. In a bar plot, the *x*-axis represents the categories and the *y*-axis represents the number of occurrences for each category. It is a useful method for comparing the various categories, as shown in Figure 4.10. Using a bar plot, the category with the highest frequency may be identified, as shown in Figure 4.10; most people like grapes.

A bar graph is also an excellent method for comparing two categorical variables, as shown in Figure 4.11. The two variables, Zoo 1 and Zoo 2, have three categories.

When it comes to the first category, Zoo 1 has a greater count than Zoo 2, but Zoo 2 has a larger count when it comes to the second and third categories. A grouped bar chart is the name given to this particular type of bar graph, which displays the many categories in a manner that places them side by side. A stacked bar chart, as shown in Figure 4.11, allows a different type of comparison. In this particular instance, the counts

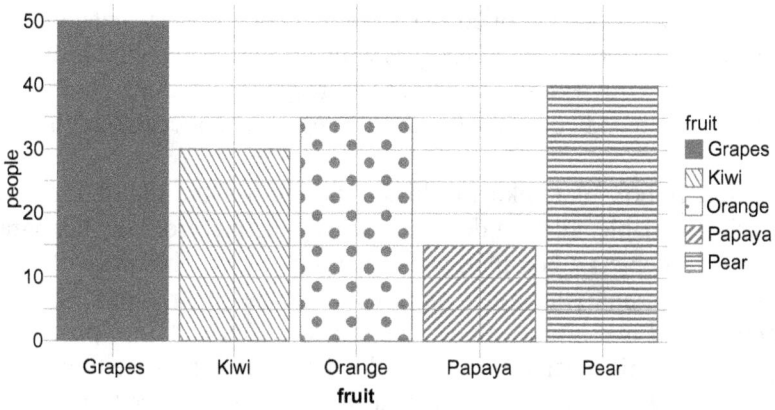

**Figure 4.10.** Bar Plot

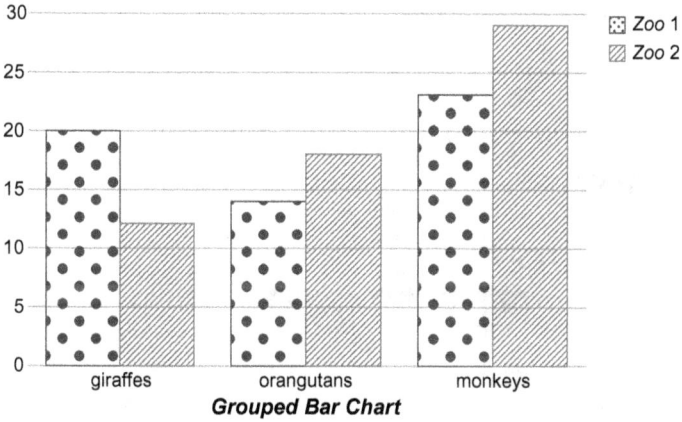

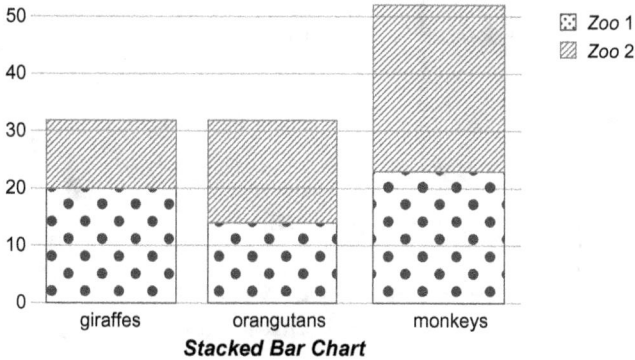

**Figure 4.11.** Bar Plot for Comparing Two Categorical Variables (Zoo 1 and Zoo 2)

for the two variables are piled over one another for every category. The combined count for the first category is virtually exactly the same as the combined count for the second category, as seen by this bar chart. On the other hand, the combined count for the third category is noticeably higher.

*Box Plot*: A box plot, like the histogram, shows the distribution of a numerical variable, albeit it does so in a way distinct from the histogram. Figure 4.12 depicts a box plot, which is used to show the range of values for a given variable (the box is represented by the grey region). The box's minimum and maximum values represent the 25th and 75th percentiles, respectively. This means that half of the data falls within the box and the other half falls outside of it, and that the median is the 50th percentile. Whiskers are two horizontal lines that represent the 10th and 90th percentiles. As a result, 80% of the information is contained inside the boundaries of the extremes.

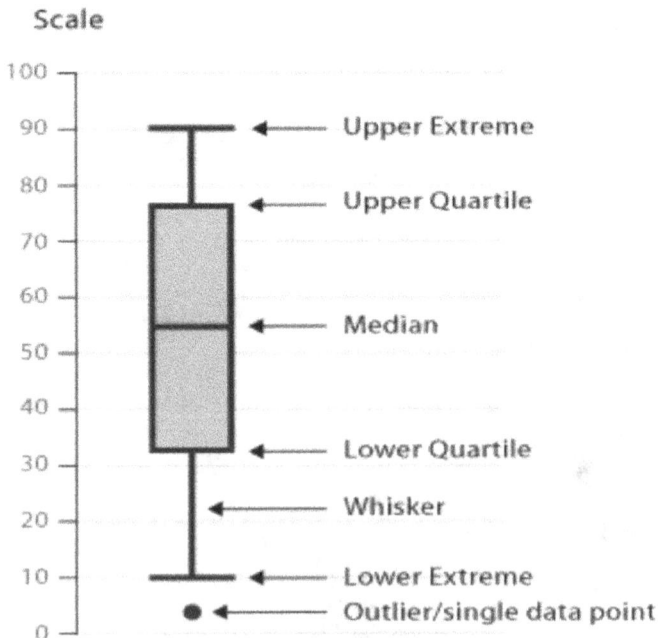

**Figure 4.12.**    Parts of a Box Plot

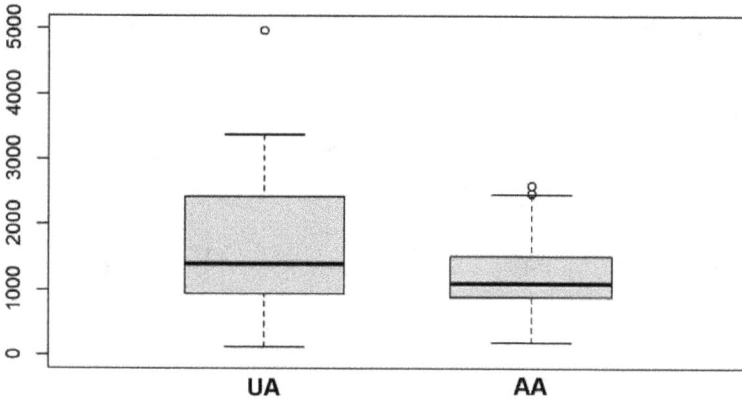

**Figure 4.13.** Box Plot of Distance Covered by UA and AA

Any data values that fall outside of the whiskers are considered outliers and are represented on the box plot by a single point. Box plots offer a condensed representation of the distribution of the variables being analyzed. As a result, they are utilized quite frequently in the process of variable comparison. The box plot given in Figure 4.13 compares the basic distance covered by United Airlines (UA) and American Airlines (AA).

The median, range, and dispersion of the two variables may all be immediately gleaned from this plot. UA has a greater median distance than AA, and it has one outlier compared to AA, which has two outliers. We can also observe more variation/spread in UA than in AA. A box plot can also reveal whether the data values are distributed symmetrically, positively skewed, or negatively skewed, as shown in Figure 4.14.

In symmetric distribution, as shown in Figure 4.14, the line in the associated box plot that depicts the median is centered within the box. This implies that the median is in the middle of the range. When the median is located to the right of the centre of the box, the skew is said to be negative. It may be deduced from this that there are a greater number of values that are less than the median than there are values that are greater than the median. In the same manner, a positive skew is denoted by the median being located to the left of the centre of the box.

### 4.2.2.2 *Pre-Processing*

The second stage of the preparation phase is called the pre-processing step. The purpose of this stage is to generate the data that will be required

Negative Skew        Symmetric        Positive Skew

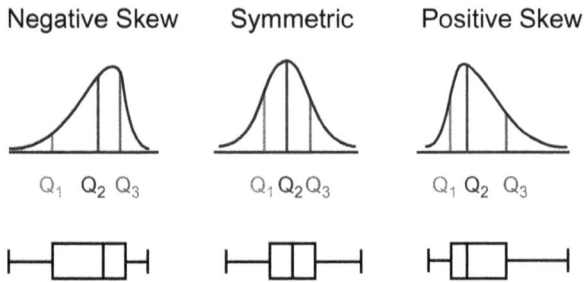

**Figure 4.14.**   Box Plot Corresponding to the Distribution Shape

for the analysis. The major purpose of this stage is to clean the data, decide which variables are appropriate to use, and convert the data, if necessary [18].

## 4.2.2.2.1 Data Cleaning

The cleaning of the data is an essential part of the data preparation process. This involves fixing problems with the data's quality, such as missing values, duplicate data, inconsistent/incorrect or invalid data, noise in the collected data that skews the true values, and outliers [19]. These issues can potentially lower data quality, which in turn might jeopardize analysis and hence the final outcomes. This is why it is crucial to identify and solve these data quality issues.

*Missing Values*: Missing data is a common problem in data quality. Each record in the dataset represents a sample with variables/features like name, age, and income. It is likely that some variables will not have any value for some of the samples that are analyzed. The term for these is "missing values" [20]. Values that are not present are indicated by the not available notation (NA). As a result, the notation NA will be used synonymously with the notation missing values. If in a data collection there is an optional field, it is possible that some of the values in that field will be blank. The age field, for instance, is sometimes an optional question in a survey. People may also opt not to disclose their income if they are uncomfortable doing so. As a result, the income variable in that dataset will have some blanks. In some circumstances, a variable might not be

**Table 4.4.** Duplicate Data

| Name | Address |
| --- | --- |
| Ram | 410/5, Yamuna Vihar |
| Jay | 611/2, Jyoti Nagar |
| Jai | 611/2, Jyoti Nagar |
| Rani | 450/3, Shanti Nagar |
| Rani | 622/4, Ram Nagar |

applicable to each and every situation. For instance, a person who is retired, jobless, or a minor might not be subject to the income requirement. Hence, there will be no record of income in any of the examples that you provide. In addition, problems with the data collection equipment, the network that carried the data, or any other factor during the data collection, transmission, or storage processes can all lead to missing values.

*Duplicate Data*: When a dataset contains data objects that are copies of one another, this results in the existence of duplicate data [21]. For example, against the same name, we have different addresses or vice versa, as shown in Table 4.4. A scenario like this could occur, for example, if the residence of a customer was updated, but their new address was added to their file rather than the updated version of their previous address being saved. When combining data from many sources, duplicates can appear.

*Inconsistent/Invalid Data*: This happens when a variable is assigned an impossible value. An example of this is a negative age or a four-digit zip code. Data input errors during data collection are a common cause of these inaccurate data values.

*Noise*: In this context, it means anything that has the potential to skew the results of your analysis. Both the data collection procedure and the data transmission process are vulnerable to the introduction of noise. For example, while recording an audio message, there may be a buzzing noise in the background owing to environmental factors or a malfunctioning microphone. Another instance is a picture that is too bright because the light exposure was adjusted wrong.

*Outliers*: A data sample with values that are significantly distinct from those of the other data samples contained inside a data collection is referred to as an outlier. Outliers can be produced whenever there is an error that leads to data that is significantly higher or lower than what is typical, and in this situation, we need to get rid of the outliers that have been produced. On the other hand, in some applications, such as the identification of fraud, outliers are the essential samples that require further study. Therefore, outliers may need to be discarded or kept for additional examination, depending on the context in which they are being used.

## 4.2.2.2.1.1 Addressing Data Quality Issues

As we know, real-world data are messy, and any further analysis performed on this data will produce misleading results. There is a need to address these data quality issues discussed in the previous section to continue further with the process of data analysis.

*Handling Missing Values*: When certain samples do not include a value for some variables, this is known as missing data. Dropping samples that include missing values or NAs is a quick and easy technique to deal with missing data. Every tool for machine learning comes with some kind of method or instruction to filter out rows that are blank or have values that are missing. This method has the benefit of being easy to implement. The downside is that when we filter out instances, we remove data, and if the number of dropped samples is high, we will lose a substantial amount of data [20].

Samples with missing data can be imputed so that they are not thrown out. Imputing is the process of substituting a set of plausible values for missing data [19,20]. It is much more involved than just ignoring the samples with missing values, but this approach lets you make use of all of your data. Many techniques exist for filling in data gaps through imputing. The average or median value of the variable can be substituted for missing data. For instance, if an employee's years of service are unknown, the median or mean of all current employees could be used as a stand-in. The most frequent value can also be substituted for the missing value. For example, the most frequently recorded age of customers associated with the specific item can be used if that value is missing. Alternatively, a sensible value can be derived as a replacement for a missing value. For consumers under the age of 18, for instance, a missing income amount can be set to zero, while for customers in other age brackets and occupations, an average number can be substituted. To replace missing values with appropriate ones, this approach requires familiarity with the program and the variable in question.

*Handling Duplicate Data*: In the case of duplicate data, deleting the older record is one option. Another method involves merging duplicate records [21]. For this reason, it's important to have a system in place for deciding how to settle values disputes. Several addresses for the same customer may necessitate the use of logic to sort through and find commonalities. For instance, St. is equivalent to Street.

*Handling Invalid Data*: To address invalid data, it may be required to consult different data sources. For instance, if the city and state are known, it may be possible to correct an incorrect zip code by conducting a search for the proper zip code. Instead, the most accurate approximation of a value that is considered reasonable can also be used in its place [21]. For instance, if an employee's age is unknown, one can reasonably estimate a value for it based on the length of time the employee has been employed by the company.

*Handling Noise*: It is possible to get rid of the noise that has the effect of changing the values of the data by filtering out its source. For instance, removing disturbance from the voice tape can be accomplished by filtering off the frequency of the background noise that is constant. Yet, extreme caution is required when carrying out this filtering because it has the ability to eliminate legitimate data components.

*Handling Outliers*: The identification of outliers is made possible through the utilization of summary statistics as well as data visualizations. The presence of outliers can significantly skew the distribution of your data and, as a result, the conclusions that you draw from your study. You will want to leave these samples out of the data-gathering process if identifying outliers is not the primary goal of your investigation. For instance, if a broken thermostat causes the readings to fluctuate unpredictably or to be much higher or lower than usual, the relevant samples must be thrown out. Nonetheless, in many different contexts, it is desirable to have outliers. Therefore, when we identify outliers, we should not discard them. We should instead analyze them more attentively [21]. A famous example of this is fraud identification, where outliers suggest the possibility of fraudulent usage, and those samples must be extensively scrutinized.

It is absolutely necessary to have a comprehensive understanding of the application in order to successfully handle any data quality issues. There should be a thorough knowledge of how the data was acquired and the application's intended use. This specialized knowledge is crucial in making well-informed decisions regarding how to effectively substitute missing values, how to manage duplicate records and incorrect data, and what to do about noise and outliers in the data.

### 4.2.2.2.2  Feature Selection/Engineering

It refers to the selection of a suitable collection of features for an application. Feature selection may entail the elimination of superfluous or unnecessary features, the combination of features, or the creation of new features [22]. During the data exploration phase, one may identify that two features are highly related. In this particular situation, the outcomes of the analysis might be unaffected even if one of these attributes was removed. It is highly likely that there is a connection between the purchase price of a product and the amount of sales tax that must be paid, for instance. Getting rid of the tax on merchandise sales will be beneficial as a result. Eliminating duplicate or unnecessary features simplifies further analysis.

The purpose of the feature selection process is to choose the smallest set of characteristics that nevertheless accurately conveys the nature of the issue being tackled. The analysis will be made easier to understand if the number of features that are considered is reduced [23]. Yet, one has to incorporate all aspects that are pertinent to the issue. So, there needs to be a compromise between how expressive the feature set may be and how much space it takes up. While selecting the features, one can add new features, delete certain features, recode features, or combine features. The feature set used for analysis is modified by these procedures.

*New Feature Addition*: Existing features can be used to develop new features [22]. For example, based on the student's state of residency, a possible feature enhancement could be to determine whether a given student is a resident of the state or not, as shown in Table 4.5.

This new element reflects an important component of a college admission application and would thus be highly useful as a distinct feature.

**Table 4.5.** Feature Addition

| Name | State | | Name | State | In State |
|------|-------|--|------|-------|----------|
| Renu | HR | | Renu | HR | T |
| Jenny | PB | | Jenny | PB | F |
| Ziva | HR | | Ziva | HR | T |
| Suraj | UK | | Suraj | UK | F |

*Feature Deletion*: Features can also be deleted; candidates for deletion include highly correlated features [22]. During data exploration, one might learn that two features are highly correlated, i.e., they vary in a highly comparable manner. For instance, there is likely to be a strong correlation between the price of a product and the total amount of sales tax paid. The sales tax increases with the purchase price. One may eliminate one of these features in this scenario, as they provide identical information. In addition, keeping both features substantially increases the size of the feature set and complicates the analysis. Due to the uncertainty surrounding the use of features with multiple missing values, features with a large percentage of missing values may also be appropriate candidates for removal. Therefore, deleting may not result in any information loss. In addition, irrelevant features should be eliminated from the data collection. A feature is irrelevant if it does not include any beneficial information for the analysis. An employee ID might be used as an illustration of this in the context of revenue projections for the employee. Additional identifiers used in the process, such as row number, person ID, and so on, are potential candidates for removal.

*Combining Features*: The combination of features is also possible if the new feature gives information that cannot be obtained by examining the original features separately [20]. The body mass index (BMI) is one measurement that can be used to determine whether or not a person is overweight. A program designed to aid in weight loss absolutely must have this component. Information about a person's weight distribution cannot be gleaned from measuring either their height or weight itself. This information can be obtained from the height and weight ratio.

*Recoding Features*: It is possible to recode a feature such that it meets the needs of the application. One scenario in which this might come into play is when an individual desires to change a continuous feature into a categorical one [22]. For the sake of a marketing application, one can decide to recode the ages of their clients into distinct demographic subsets, such as adolescents, young adults, adults, and senior citizens. Teenagers are considered to be those aged 12 to 18, young adults are those aged 19 to 24, adults are those aged 25 to 58, and senior people are those aged 58 and more. A further instance of feature recoding may be coding a feature to indicate whether a consumer prefers to purchase costly things. Here, we could rewrite the code such that it returns 1 if the customer's average order value is greater than some threshold and 0 otherwise. As features are recoded, they are sometimes split up into many features. One typical use of "breaking features" is to disassemble a full address into its component parts, such as "street address", "city", and "state". It allows us to quickly arrange records by state, for example, to analyze data statewise.

### 4.2.2.2.3  Feature Transformation

Data are transformed from one format into another during the feature transformation phase. There are several transformation operations, including scaling, aggregation, and dimensionality reduction 6 [24].

*Scaling*: Scaling, as shown in Figure 4.15, maps input data to a specified range so as to prevent any one feature from dominating the conclusions of the investigation.

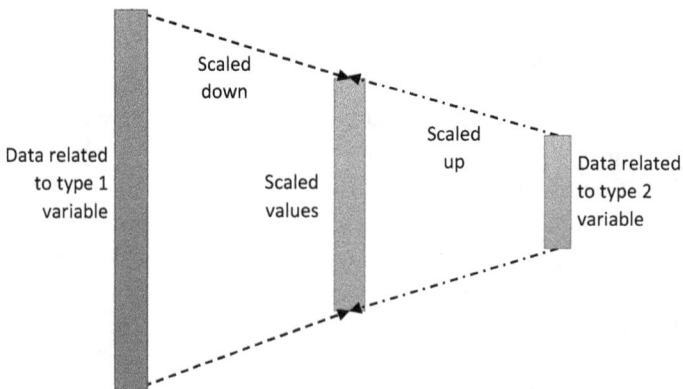

**Figure 4.15.**    Scaling Transformation

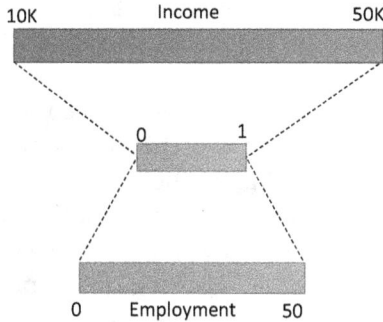

**Figure 4.16.**   Scaling to a Range between 0 and 1

Mapping all values of a feature to a given range, such as between zero and one, is one technique to conduct scaling. For instance, suppose there is a feature for income with a range from ₹30,000 to ₹100,000 and another variable for employment in years ranging from 0 to 50. These features have vastly distinct scales. To equally weight these features for comparing data samples, the range of both features can be scaled between 0 and 1, as shown in Figure 4.16. This will prevent the income variable from dominating the comparison result.

Alternately, scaling can be accomplished by transforming the features so that the results have a mean of zero and a standard deviation of one. Calculating the mean ($\mu$) and standard deviation ($\sigma$) for the feature to be scaled is the initial step in the scaling process. Subtract the mean value from each value for this feature and divide by the standard deviation $\left(\frac{x_i - \mu}{\sigma}\right)$. The transformed feature will have a mean of zero and a standard deviation of one, effectively removing units of the features. This scaling method is called standardization or normalization.

*Filteration*: Filteration is another feature transformation method that minimizes data noise and variability. Typically, it is applied to time-series data, such as speech or audio signals. The noise in speech or audio signals is generally a high-frequency component, which can be eliminated using a low-pass filter. Filtering can also be used to eliminate noise in an image. Noise is a random change in the intensity or color of an image's pixels. A noise, for instance, might cause a picture to look grainy. A mean or median filter can be used to replace the value of a pixel with the mean or median of its nearby pixels. This has the effect of smoothing down the image by eliminating the noise responsible for its graininess [25].

*Aggregation*: The feature values are aggregated to summarize data or minimize variance. Aggregation is performed by summing or averaging data values at a higher level. Aggregation can eliminate noise to create a clear representation of the structure of the data [26]. For instance, hourly values can be aggregated to the daily level, or data within a city area might be aggregated at the state level. Tracking stock prices is an example of aggregation. A stock's hourly variances may be challenging to observe, but the stock's daily fluctuations may better highlight any upward or downward trend.

*Dimensionality Reduction*: It helps in reducing the number of data dimensions to facilitate further analysis. Frequently, the data utilized in machine learning procedures contain several variables; such data are referred to as high-dimensional data [27]. Most of these parameters may or may not be relevant to our application in light of the issues we pose. The effectiveness and correctness of the analysis can be improved by paring down the dimensions to a more reasonable number of valuable and related variables. The number of features or variables in a data collection serves as the primary determinant of the number of dimensions or dimensionality of the collection. If the dataset has two features, then it is two-dimensional and similarly for subsequent numbers of features. To capture the characteristics of data, we wish to incorporate as many features as possible without increasing the dimensionality of the data. As dimension rises, problem spaces increase exponentially, and data become scarcer as space expands. Figure 4.17 shows how the problem space increases from one dimension to two and three dimensions. The quality of the analysis suffers when the dataset is small in contrast to the overall problem space. In addition, it becomes significantly more complex to specify and compute key computations employed in the analysis. The aforementioned challenges are examples of how difficult it may be to work with high-dimensional data, which is why this problem is also referred to as the "curse of dimensionality". We ought to decrease the dimensionality of the data so that we can break free from this curse. At this step, one will narrow down the number of features to a subset that is more likely to represent the characteristics of your data accurately.

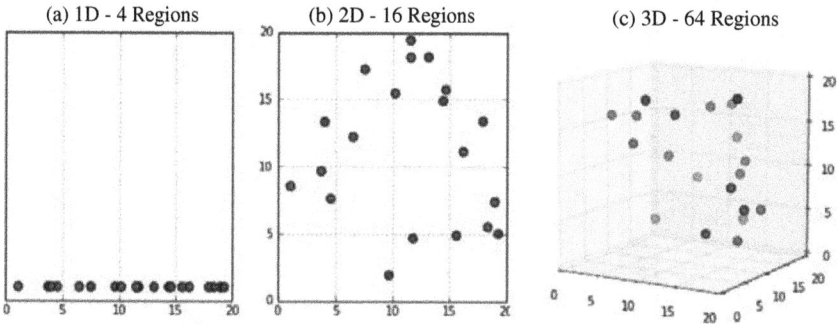

**Figure 4.17.** Effect of Increasing Dimensionality of Data

One approach to reduce the dimensionality is to remove features that are correlated with other features. Furthermore, another technique for decreasing the number of dimensions is to use mathematics to select the dimension that is the most important to keep and to disregard the other dimensions. The goal is to narrow down the number of dimensions until we find the optimal combination that accounts for the most variations in the data. This simplifies further analysis by reducing the data's dimensionality and removing irrelevant features. Principal component analysis (PCA) [28] is a technique widely used to identify the subset of the most essential dimensions. PCA aims to reduce the number of dimensions used to store data while maintaining the same level of detail. Basically, PCA seeks to determine which selection of dimensions most effectively summarizes the data.

Figure 4.18 illustrates the idea behind PCA. The data samples are plotted in a 2D space defined by *x*-axis and *y*-axis. Most of the scatter in the data occurs along the red diagonal line. This suggests that this is the dimension along which the data samples can be most easily differentiated. The red line represents the first fundamental dimension, PC1. It gives a numerical value to the vast quantity of informational diversity along a selected axis. The red line represents PC1, which is not perpendicular to either axis. To find the principal component that captures the next largest amount of data variance, one must turn one's gaze in the orthogonal, or perpendicular, direction to the first principal component. The second primary component, PC2, is depicted by the green line in the graph. This procedure can be repeated until all desired primary components have

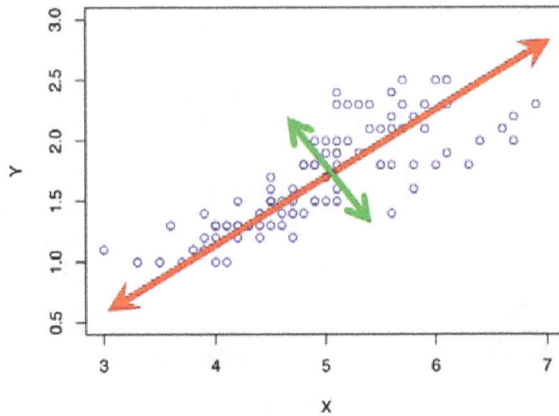

**Figure 4.18.**   Principal Components Analysis

been identified. Neither the *x*-axis nor the *y*-axis is aligned with the primary components. Moreover, they are orthogonal or perpendicular to one another. This is the role of PCA. It identifies the highlighted dimensions or the key components that account for as much data variance as feasible. In place of standard dimensions such as *X*, *Y*, and *Z*, these principal components constitute a new coordinate system for transforming data.

But the question is: How does PCA aid in dimension reduction? Let's take another look at the PC1 component shown by the long diagonal line in Figure 4.18. Since the first principal component accounts for the greater part of the variability in the data, the initial data sample may be mapped to the dark-lined dimension with just a small amount of information being lost in the process. In this particular scenario, we mapped a two-dimensional dataset to a one-dimensional space while preserving as much of the information as we could. The following are a few of the most important aspects of PCA. The PCA will establish new coordinates for the data in such a way that the first coordinate defined by the first principal component will correspond to the new coordinate system. It captures the largest diversity in the data. In a dataset, the second coordinate, which is defined by the second principal component, represents the next highest variance, etc. You can establish a lower-dimensional space for the data by using the first few principal components, which are the components that account for the majority of the variance in the data. When dealing with high-dimensional data, PCA can be an extremely helpful method for reducing the data's dimension. While PCA is a beneficial technique for reducing the dimensionality of the data, it has the potential to make the analysis models that it generates more challenging to interpret. The initial features of a dataset have their own unique significance. However, the transformed/modified data have dimensions that no longer have their natural meanings because these are mapped to a new coordinate system that was established by primary components.

## 4.2.3 Analyze

The process of machine learning continues with the next step, which is data analysis. Constructing a machine learning model, conducting an analysis of the data, and evaluating the effectiveness of the model are the goals of this phase [29]. The analysis phase begins with determining the problem type and selecting relevant machine learning algorithms (classification, regression, association analysis, cluster analysis) to analyze data. Once the model has been constructed, it is applied to new data points to assess its performance.

### 4.2.3.1 *Classification*

In the classification problem, the machine learning model is supplied with the input data, and the goal is to make a prediction of the target that corresponds to the data that was input. Provided that the target is a categorical variable, the classification problem entails making an educated guess as to the category or label that the target falls under on the basis of the input data. For instance, the classification task depicted in Figure 4.19 involves predicting the weather. Prediction of weather is the objective that the model needs to be able to anticipate, and its potential values are cloudy, sunny, windy, and rainy. The input data may consist of measurements, such as temperature, humidity levels, air pressure, the direction and speed of the wind, and other similar variables. As a result, the purpose of the model is to make a prediction of whether or not the weather for the day will be cloudy, windy, sunny, or any combination of these four conditions given specified values for temperature, relative humidity, and atmospheric pressure.

The dataset for the weather classification problem is shown in Table 4.6.

Every table row represents a sample, with the variables such as temperature, humidity, and pressure serving as inputs and the variable

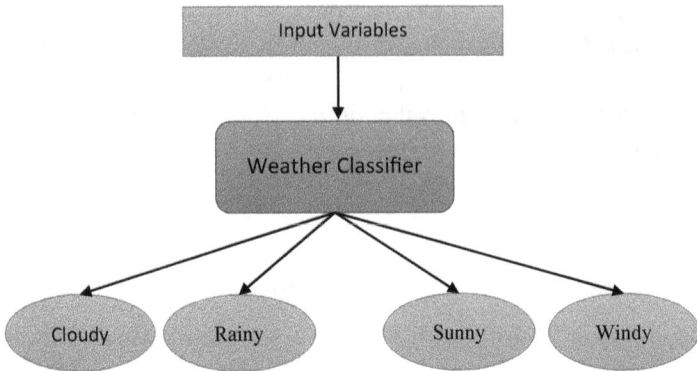

**Figure 4.19.**    Weather Classifier

**Table 4.6.**    Dataset for Weather Classifier

| Input Variables | | | Target |
|---|---|---|---|
| Temperature | Humidity | Pressure | Weather |
| 70 | 45 | 1020.1 | Sunny |
| 65 | 80 | 1010.6 | Rainy |
| 68 | 48 | 1018.7 | Windy |
| 55 | 77 | 1086.3 | Cloudy |

weather serving as the target. There are a number of names that can be used to refer to the target variable, including target, label, output, class variable, category, and class. The classification task consists of making an educated guess as to what the value of the target variable will be based on the values of the other variables. We are given a target or labelled data, therefore classification is a supervised task.

There are two possible types of classification problems: binary and multi-class. In binary classification, the value of the target variable can only take on one of the two possible forms, including yes or no. In multi-class classification, the target variable can take on more than two different values at the same time. Many people often refer to multi-class classification as multinomial classification or multi-label classification. The practice of attempting to forecast whether or not it will rain the following day is an illustration of binary classification, in which there are only two possible outcomes: yes, which indicates that it will rain the following day, or no, which indicates that it will not rain the following day.

Another example of binary classification is determining whether or not a transaction made with a credit card is legitimate or fraudulent. Predicting the type of product a consumer would purchase is an example of multi-class categorization. Product categories such as kitchen, electronics, and clothing would serve as potential values for target variables. Another example of multinomial classification is classifying the emotion of a tweet as favorable, negative, or neutral.

## 4.2.3.1.1. Building and Applying a Classification Model

A machine learning model as shown in Figure 4.20 is represented as a mathematical model. This model involves parameters and makes use of some mathematical formulas in order to determine the connection that exists between inputs and outputs.

The model uses the parameters to adjust the inputs in order to create the outputs. This input–output relationship is corrected or refined by the model's parameter adjustments. The parameters of a machine learning model are tuned or approximated using data in a learning algorithm. This procedure is also known as creating a model, building a model, training a model, or fitting a model.

In general, there are two steps involved in the construction of classification models or any other machine learning models. The training phase is the first step, and it is during this phase that the model is formed and its parameters are changed by making use of the training data. The data utilized to construct or "train" a model are referred to as "training data". The testing phase is the second step. The process of applying the learned model to new data, which refers to data that were not used in the training of the model. The model has never had the opportunity to look at the test data, which are distinct from the training data. The performance of the model is then evaluated based on the test data.

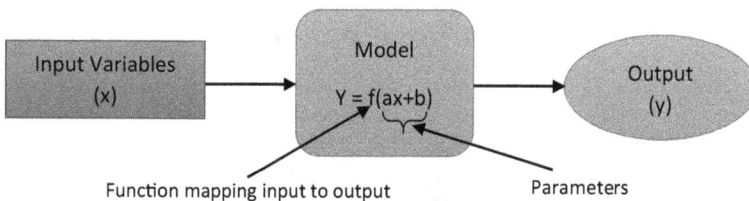

**Figure 4.20.** Machine Learning Model

## 4.2.3.1.2  Classification Algorithms

A classification model modifies its parameters such that its outputs correspond to the desired outcomes. To update the parameters of a model, a learning method is utilized. There are several classification model-building techniques, including $k$-NN or $k$-Nearest Neighbors, decision tree, and naïve Bayes [30].

$k$-*NN*: $k$-Nearest Neighbor is the simplest technique for building classification models. The key idea is to categorize a sample according to its neighbors. For a newly arrived sample, the label of its class is determined by looking at the labels of the samples that are immediately surrounding it, as depicted by the green circle in Figure 4.21. The duck test is the foundation upon which $k$-NN is built. If it acts like a duck and makes duck noises, then it is probably a duck.

In the context of classification, this indicates that samples with comparable input are most likely part of a single class. As a consequence of this, samples that have identical values for their inputs ought to be labeled using the identical target label. This suggests that the classification of a sample is dependent on the target labels of the points that are immediately adjacent to it.

Given a new sample shown as the shaded circle in Figure 4.22, the aim is to identify the training data samples that are most similar to the new sample, in other words, we can say, determining the label for the new

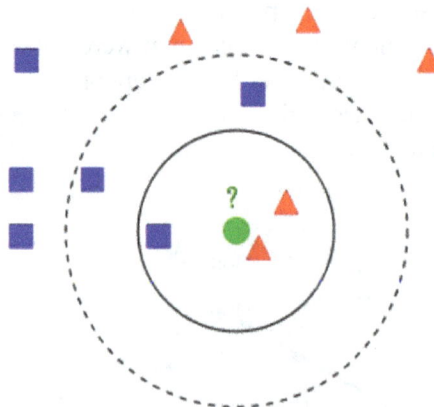

**Figure 4.21.**   $k$-NN Classification

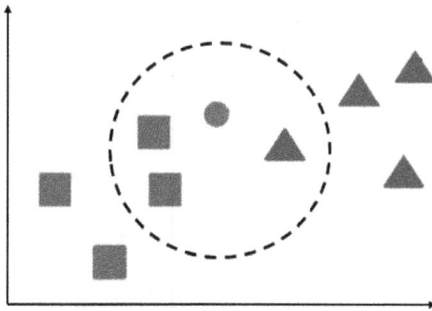

**Figure 4.22.**   *k*-NN Working

sample based on the labels of the nearby points. In Figure 4.22, it is equivalent to determining whether this sample should be categorized as a shaded square or a shaded triangle. The value of $k$ determines how many neighbors are considered in determining the label of the new sample. If $k = 1$, just the nearest neighbor is evaluated to determine the class of the new sample, if $k = 2$, the two closest neighbors are considered, and so on. While considering multiple neighbors, a voting system is employed. Majority of the vote is usually employed, thus the label connected with the majority of the neighbors is applied to the new sample. With $k$-NN, a measure of similarity is required to assess the proximity of two samples. This is required in order to identify which samples are nearest neighbors. The most commonly used distance measures are Euclidean distance, Manhattan distance and Hamming distance. In conclusion, $k$-NN is a straightforward classification method. Notably, there is no dedicated "training" phase where the actual model is created, and its settings are fine-tuned. However, it can be susceptible to noise because classification decisions depend only on a small number of surrounding points rather than the entire dataset. Finding $k$-NN can be time-consuming because the distance between a new sample and every sample point in the data must be calculated.

*Decision Trees*: It is a popular algorithm for classification. The idea of decision trees for classification is to divide the data into subsets that belong to just one class [31]. This is achieved by splitting the input space into pure areas or regions containing only samples from one class. With actual data, it may be impossible to create absolutely pure subsets.

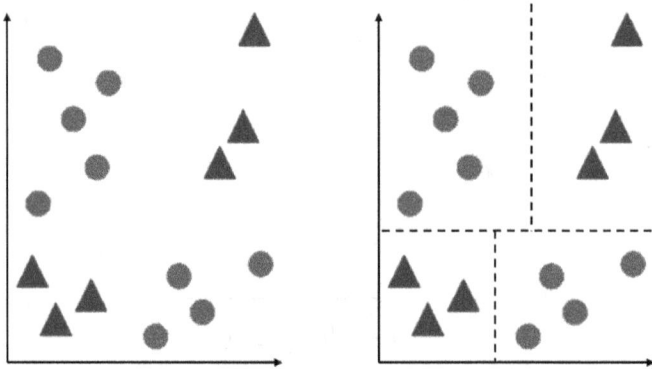

**Figure 4.23.** Decision Tree Classification

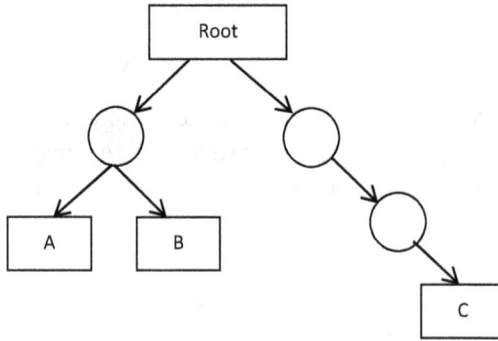

**Figure 4.24.** Structure of a Decision Tree

Therefore, the objective is to split the data into as pure subsets as feasible. That is, each subset contains as many instances of a single class as feasible, as shown in Figure 4.23.

A decision tree is a structured hierarchy containing nodes and directed edges, as shown in Figure 4.24. The topmost node is known as the root node. The nodes at the bottom are known as leaf nodes, while the remaining nodes are known as internal nodes. There are test criteria connected with the root and internal nodes, and each leaf node has a class label. A categorization determination is reached by traversing the decision tree from the root node outward. At each node, the response to a test condition selects the branch to follow. When a leaf node is reached, the classification decision is determined by the category at the leaf node.

The depth of a node is the number of edges connecting it to the root node. Zero is the depth of the root node. The depth of a decision tree is the number of edges in the longest path between the root and leaf nodes. The number of nodes in a decision tree determines its size.

The technique used to generate a decision tree model is known as an induction algorithm. At each split, the induction algorithm only evaluates the optimal option to split the data at hand. This is known as a greedy strategy. Given a data collection, it is not possible to know the optimal tree, hence the tree must be constructed piecemeal by calculating the optimal method to divide the current node at each stage and integrating these decisions to produce the final decision tree. How does a decision tree identify the optimal method to divide a sample set at a node? The objective is to divide data at a node into as pure subsets as feasible. The Gini Index is a popular impurity metric used to determine the optimal split. The higher the purity of the divide, the lower the Gini Index. Therefore, the decision tree will choose the partition that minimizes the Gini Index. Other measurements of impurity include entropy, information gain, and misclassification rate in addition to the Gini Index.

*Naïve Bayes*: In a naïve Bayes classification model, the relationships between the input features and the class are given as probabilities. This is a probabilistic method of classification. Therefore, the probability for each class is estimated given the input attributes for a sample. The sample's label is then chosen according to the class with the highest probability. The naïve Bayes classifier applies Bayes' theorem in addition to a probabilistic framework for classification. Estimation of the probability is made simpler by the use of Bayes' theorem. Further, the input features are also presumed to be statistically independent of one another. This independence assumption is viewed as naïve because it is oversimplified and does not always hold true. This classification model gets its name from the Bayes theorem and the naïve independence assumption.

Let $X = \{X_1, X_2,..., X_n\}$ denote the input feature. Given a sample with features $X$, we have to predict the class $C$, and this is done by finding the value of $C$ that maximizes $P(C|X)$. We must determine the conditional probability of class $C$, given $X$ for all classes, i.e., find $P(C_1|X)$, $P(C_2|X)$, ...., $P(C_n|X)$, and choose the class with the highest probability in order to determine the class label $C$. However, it is challenging to estimate this probability since we would need to list every conceivable combination of feature values to know the conditional probability. This is where the

Bayes Theorem comes into picture. The Bayes theorem can be used to rephrase the classification problem, making it easier to understand.

According to the Bayes Theorem,

$$P(C \mid X) = \frac{P(X \mid C) \times P(C)}{P(X)}$$

where $P(C|X)$ is referred to as posterior probability. Since $P(C|X)$ represents the likelihood that the class label will be $C$ after looking at input features $X$. $P(X|C)$ is the likelihood that input features $X$ will be observed given that $C$ is the class label. Since it depends on the class, this is the class conditional probability. $P(C)$ is the likelihood that a class label, such as "C", actually exists before any input data have been observed. Consequently, it is known as the prior probability. $P(X)$ is the likelihood that input feature $X$ will be observed, regardless of the class label. We therefore wish to determine the posterior probability $P(C|X)$ for each class $C$ in the classification process. Given the input $X$, the probability of $X$ is a constant because it is independent of the class $C$. Since it is the same for all classes, $P(X)$ can be eliminated from the $P(C|X)$ computation. Hence, $P(C|X) = P(X|C) \times P(C)$, which can be calculated from the data.

### 4.2.4 Evaluation of Machine Learning Models

Machine learning model translates the input received into an output. The target is the actual class label, while the outcome of a classifier is the predicted class label based on the input data. Success for a classifier is defined by its ability to anticipate a sample's class label accurately. If the actual class label does not match the predicted class label, an error occurs. As a result, the error rate is the total percentage of errors in the data. Simply said, it is the ratio of the number of incorrectly classified samples to the total number of samples. The misclassification rate is synonymous with the error rate. The error rate on the training data is known as training error, whereas the error rate on the test data is known as test error. Generalization [32] measures how well a classifier performs on novel data by looking at the error in the test data. The extent to which a model retains its predictive ability when applied to data that was not used to train it is referred to as generalization. If the model generalizes successfully, it will be able to produce accurate results when applied to datasets that share

similar structures with the training data but do not contain the same samples. The term "generalization error" also refers to "test error", as it
reveals how effectively the model generalizes to new data [32].

The concept of overfitting is connected to that of generalization.
Overfitting occurs when a model has exceptionally low levels of error
during training but significant levels of error during generalization [32].
This suggests that the model has been trained to mimic the noise present
in the training data rather than the fundamental structure of the data itself,
as shown in Figure 4.25.

As can be seen in Figure 4.25, the mapping of input to output learnt
by the system is shown in the form of an arc, while training data are
shown in the form of points. Since the curve follows the sample point's
trend, the figure on the left provides evidence that the model has acquired
an understanding of the fundamental structure of the data. Yet, as shown
by the graph on the right, the model has picked up the skills necessary to
recreate the noise that is present in the dataset. Instead of focusing on the
overall pattern of the samples, the model makes an effort to account for
each individual sample point. The term "poor generalization" can also be
used to refer to overfitting. An overfitting model will not generalize well
on fresh or unknown data. As a consequence of this, the model will only
perform well when supplied with the data on which it was trained; when
presented with new data, it will perform very poorly. Underfitting is
another problem similar to overfitting, as shown in Figure 4.26.

Overfitting occurs when the model fits the noise that is present in the
data used for training. Due to this, the training error is minimal, while the

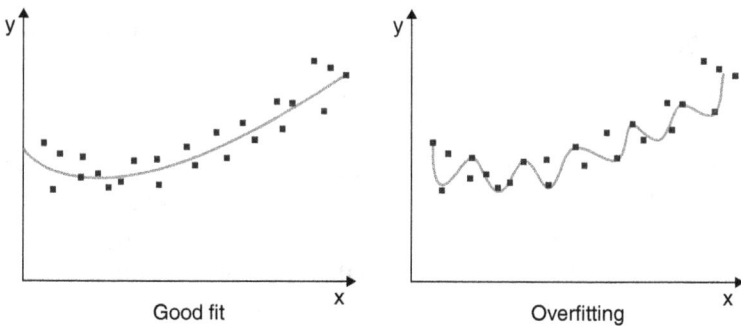

**Figure 4.25.** Overfitting Issue in Machine Learning Model

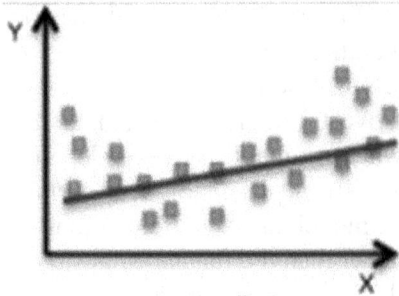

**Figure 4.26.** Underfitting Issue in Machine Learning Model

**Table 4.7.** Types of Classification Errors

| True Label | Predicted Label | Error Type |
| --- | --- | --- |
| Yes | Yes | True Positive (TP) |
| No | No | True Negative (TN) |
| No | Yes | False Positive (FP) |
| Yes | No | False Negative (FN) |

test error is significant. In contrast, underfitting occurs when the model does not adequately capture the structure of the data. As a result, both the training and test errors increase. These are problematic since they suggest that the model is overspecific.

### 4.2.4.1 *Evaluation Metrics*

This section discusses the different metrics used to evaluate the performance of a machine learning model.

When the predicted class label does not match the observed class label, an error has been made in the classification task [33]. Consider a binary classification problem, where we want to predict if the animal is a cat or not. The two class labels will be "yes" indicating that the animal is a cat or "no" indicating that the animal is not a cat. Now, depending upon the true and the predicted label, we could have the possibilities, as shown in Table 4.7.

These classification errors can be represented in the form of a matrix which is called as the confusion matrix, as shown in Figure 4.27.

| | | Predicted Class Label | |
|---|---|---|---|
| | | Yes | No |
| True Class Label | Yes | TP | FN |
| | No | FP | TN |

**Figure 4.27.** Confusion Matrix for Binary Classification Problem

In computing various classifier assessment metrics, these four kinds of errors play a vital role. The rate of correctness, also known as accuracy, is the most common evaluation statistic employed. Accuracy in classification is determined by dividing the number of right predictions by the total number of predictions:

$$\text{Accuracy Rate} = \frac{\#\text{correct predictions}}{\#\text{total predictions}} = \frac{TP + TN}{TP + TN + FP + FN}$$

There is another measure called error rate, which can also be used to express the model's performance. The rate of errors is the reverse of accuracy.

$$\text{Error Rate} = \frac{\#\text{incorrect predictions}}{\#\text{total predictions}} = \frac{FP + FN}{TP + TN + FP + FN}$$
$$= 1 - \text{Accuracy Rate}$$

When there is a class imbalance issue, accuracy and error rates give misleading interpretations. This occurs when there are a small number of instances of the class of interest and the bulk of the examples are unfavorable. This is illustrated by determining if a tumor is malignant or not. Identifying samples with cancerous tumors is of interest, however instances in which the tumor is cancerous are extremely rare. That's why the bulk of samples are negative, and the positive samples make up a small percentage. Therefore, the term "class imbalance" is used to describe the issue. If accuracy is used in the situation of class imbalance, then what could be the issue? Consider a scenario in which only 4% of the instances involve malignant tumors. If the classification model always predicts non-cancerous tumors, then its accuracy rate will be 96%, as 96% of the samples will have non-cancerous tumors. This is due to the fact that 97% of the samples will contain cancerous tumors. It is important to note, however, that the model does not uncover any instances of cancer in this particular scenario. As a consequence of this, the accuracy rate is quite misleading. With the great level of accuracy it displays, we could be led to conclude that your model is operating quite effectively. Nevertheless, it was not possible to locate any cases belonging to the class that was being investigated. In these situations, we need evaluation metrics that evaluate the accuracy of the model in categorizing positive classifications as opposed to negative classifications.

Precision and recall are two prominent evaluation metrics used when there is an imbalance class scenario. The term "precision" refers to the ratio of "true positives" to "total positives", which includes both "real" and "false" results. In other words, it refers to the percentage of true positives in relation to the total number of samples expected to be positive:

$$\text{Precision} = \frac{\#\,\text{true positives}}{\text{All samples with Predicted} = \text{Yes}} = \frac{TP}{TP + FP}$$

The definition of recall is the ratio of true positives divided by total positive and negative results. It is the ratio of the total number of samples that belong to the true class to the total number of samples that are considered true.

$$\text{Recall} = \frac{\#\,\text{true positives}}{\text{All samples with True} = \text{Yes}} = \frac{TP}{TP + FN}$$

The term "precision" refers to a measure of "preciseness" because it determines the percentage of expected positive samples that really fall into the positive category. Since it determines the percentage of positive samples that have been accurately detected by the model, recall is commonly regarded as a measure of completeness.

The $F$-measure is a single metric that combines precision and recall, which is defined as follows:

$$F_1 = 2 * \frac{\text{Precision} * \text{Recall}}{\text{Precision} + \text{Recall}}$$

There are a variety of *F*-measure variants. The $F_1$ measure, which is the most common variant of the $F$ measure, is represented by the above equation. $F_1$ ranges from 0 to 1, with higher values giving better performance. Precision and recall are equally weighted on the $F_1$ scale. The $F_2$ measure prioritizes recall over precision, and the $F_{0.5}$ measure prioritizes precision over recall.

## 4.3 Scaling Up Machine Learning Algorithms

This section discusses the significance of distributed computing platforms such as Hadoop and Spark in the process of applying machine learning techniques to large amounts of data as well as describing how machine learning techniques can be applied to large amounts of data. Scaling up can be accomplished in a number of ways, one of which is by augmenting the capabilities of the existing systems by adding more memory, processors, and storage space so that more data can be stored and processed [34]. However, this is not the big data approach. Another approach could be using special hardware units like graphical processing units (GPU) to increase the processing capabilities for the massive data. Even though it is a good approach, the big data approach does not utilize this method. Since it is quite costly, it will eventually hit a limit. An alternative approach is to scale out, where the idea is to use a distributed system, where data are spread out among all of these systems so that they may be processed more quickly, as shown in Figure 4.28.

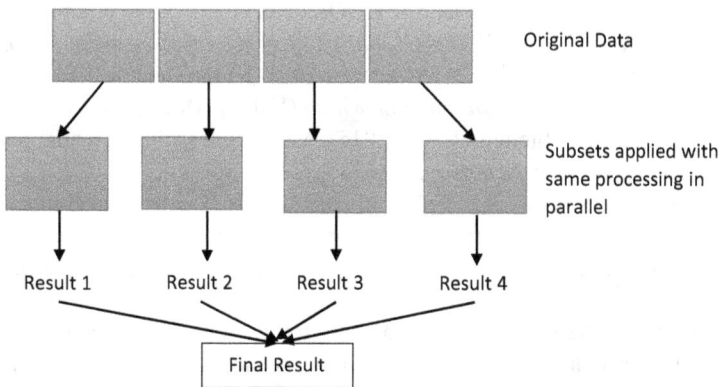

**Figure 4.28.** Distributed Approach for Processing Data

Scalable machine learning algorithms use the same scale-out methodology that was covered earlier in this section while operating in a distributed computing environment, such as Spark or Hadoop. The data are split up and distributed among multiple processors, each of which applies map, reduce, and other distributed parallel transformations to subsets of the data in parallel. This enables machine learning techniques to be applied to extremely large volumes of data.

## 4.4  Chapter Summary

In this chapter, we delved into the machine learning process, explored and prepared data, addressed the many types of machine learning tasks, examined metrics and methods for testing a model, and learned how to employ scalable machine learning algorithms for big data challenges. Further, we will again touch upon these basic topics while doing a hands-on analysis, in Chapter 10, which will include exploring data, handling missing values, and classification and evaluation with KNIME and SPARK.

## References

1. Stergiou, C. L., Plageras, A. P., Psannis, K. E., & Gupta, B. B. (2020). Secure machine learning scenario from big data in cloud computing via Internet of Things network. In *Handbook of Computer Networks and Cyber Security: Principles and Paradigms* (pp. 525–554). Springer Cham, Springer Nature, Switzerland.
2. Gupta, B. B., Gaurav, A., & Peraković, D. (2021, October). A big data and deep learning based approach for DDOS detection in cloud computing environment. In *2021 IEEE 10th Global Conference on Consumer Electronics (GCCE)* (pp. 287–290), 12–15 October 2021, Kyoto, Japan. IEEE.
3. Mitchell, T. M. (2007). *Machine Learning* (Vol. 1). McGraw-Hill, New York.
4. Jordan, M. I. & Mitchell, T. M. (2015). Machine learning: Trends, perspectives, and prospects. *Science*, 349(6245), 255–260.
5. Prakash, C., Kumar, R., & Mittal, N. (2018). Recent developments in human gait research: Parameters, approaches, applications, machine learning techniques, datasets and challenges. *Artificial Intelligence Review*, 49, 1–40.
6. Syam, N. & Sharma, A. (2018). Waiting for a sales renaissance in the fourth industrial revolution: Machine learning and artificial intelligence in sales research and practice. *Industrial Marketing Management*, 69, 135–146.

7. Kitanaka, H., Kwiatek, P., & Panagopoulos, N. G. (2021). Introducing a new, machine learning process, and online tools for conducting sales literature reviews: An application to the forty years of JPSSM. *Journal of Personal Selling & Sales Management*, 41(4), 351–368.

8. Singh, Y., Bhatia, P. K., & Sangwan, O. (2007). A review of studies on machine learning techniques. *International Journal of Computer Science and Security*, 1(1), 70–84.

9. Soofi, A. A. & Awan, A. (2017). Classification techniques in machine learning: Applications and issues. *Journal of Basic & Applied Sciences*, 13, 459–465.

10. Uysal, I. & Güvenir, H. A. (1999). An overview of regression techniques for knowledge discovery. *The Knowledge Engineering Review*, 14(4), 319–340.

11. Saxena, A., Prasad, M., Gupta, A., Bharill, N., Patel, O. P., Tiwari, A., Joo, E. M., Weiping, D., Lin, C. T. (2017). A review of clustering techniques and developments. *Neurocomputing*, 267, 664–681.

12. Tan, P. N., Kumar, V., & Srivastava, J. (2004). Selecting the right objective measure for association analysis. *Information Systems*, 29(4), 293–313.

13. Alloghani, M., Al-Jumeily, D., Mustafina, J., Hussain, A., & Aljaaf, A. J. (2020). A systematic review on supervised and unsupervised machine learning algorithms for data science. In *Supervised and Unsupervised Learning for Data Science* (pp. 3–21). Springer Cham, Springer Nature, Switzerland.

14. Michalski, R. S., Carbonell, J. G., & Mitchell, T. M. (1983). An overview of machine learning. *Machine Learning*, 3–23. Springer-Verlag, Berlin, Heidelberg GmbH.

15. Tukey, J. W. (1977). *Exploratory Data Analysis* (Vol. 2, pp. 131–160), Addison-Wesley, Massachusetts.

16. Gupta A. K. (2022) Analysis of exploratory data analysis tools and techniques. *Data Science Insights Magazine*, (Vol. 2, pp. 13–16). Insights2 Techinfo.

17. Nasser, A., Hamad, D., & Nasr, C. (2006, April). Visualization methods for exploratory data analysis. In *2006 2nd International Conference on Information & Communication Technologies* (Vol. 1, pp. 1379–1384), 24–28 April 2006, Damascus, Syria. IEEE.

18. http://robslink.com/SAS/democd18/temperb.htm.

19. Van den Broeck, J., Argeseanu Cunningham, S., Eeckels, R., & Herbst, K. (2005). Data cleaning: Detecting, diagnosing, and editing data abnormalities. *PLoS Medicine*, 2(10), e267.

20. Rahm, E. & Do, H. H. (2000). Data cleaning: Problems and current approaches. *IEEE Data Engineering Bulletin*, 23(4), 3–13.

21. Chu, X., Ilyas, I. F., Krishnan, S., & Wang, J. (2016, June). Data cleaning: Overview and emerging challenges. In *Proceedings of the 2016 International Conference on Management of Data* (pp. 2201–2206).
22. Li, J., Cheng, K., Wang, S., Morstatter, F., Trevino, R. P., Tang, J., & Liu, H. (2017). Feature selection: A data perspective. *ACM Computing Surveys (CSUR)*, 50(6), 1–45.
23. Wang, Y., Vijayakumar, P., Gupta, B. B., Alhalabi, W., & Sivaraman, A. (2022). An improved entity recognition approach to cyber-social knowledge provision of intellectual property using a CRF-LSTM model. *Pattern Recognition Letters*, 163, 145–151.
24. Tian, J. & Kang, M. (2018). Machine learning: Data pre-processing. In *Prognostics and Health Management of Electronics: Fundamentals, Machine Learning, and the Internet of Things* (pp. 111–130). Wiley-IEEE Press, John Wiley and Sons Ltd.
25. Elmisery, A. M., Sertovic, M., & Gupta, B. B. (2017). Cognitive privacy middleware for deep learning mashup in environmental IoT. *IEEE Access*, 6, 8029–8041.
26. Li, D., Deng, L., Gupta, B. B., Wang, H., & Choi, C. (2019). A novel CNN based security guaranteed image watermarking generation scenario for smart city applications. *Information Sciences*, 479, 432–447.
27. Van Der Maaten, L., Postma, E., & Van den Herik, J. (2009). Dimensionality reduction: A comparative. *Journal of Machine Learning Research*, 10(66–71), 13.
28. Abdi, H. & Williams, L. J. (2010). Principal component analysis. *Wiley Interdisciplinary Reviews: Computational Statistics*, 2(4), 433–459.
29. Mitchell, T. M. (1999). Machine learning and data mining. *Communications of the ACM*, 42(11), 30–36.
30. Aggarwal, C. C. & Zhai, C. (2012). A survey of text classification algorithms. *Mining Text Data*, 163–222. DOI: 10.1007/978-1-4614-3223-4_6.
31. Priyam, A., Abhijeeta, G. R., Rathee, A., & Srivastava, S. (2013). Comparative analysis of decision tree classification algorithms. *International Journal of Current Engineering and Technology*, 3(2), 334–337.
32. Das, S., Dey, A., Pal, A., & Roy, N. (2015). Applications of artificial intelligence in machine learning: Review and prospect. *International Journal of Computer Applications*, 115(9), 31–41.
33. Tharwat, A. (2021). Classification assessment methods. *Applied Computing and Informatics*, 17(1), 168–192.
34. Bekkerman, R., Bilenko, M., & Langford, J. (Eds.). (2011). *Scaling Up Machine Learning: Parallel and Distributed Approaches*. Cambridge University Press, New York, NY, USA.

# Chapter 5

# Big Data Analytics Through Visualization

Graphs for data analysis have a wide range of applications in both academia and industry. They can be used to model parts of human knowledge and are called *semantic networks*. Graphs dominate all the technology giants, such as Facebook, LinkedIn, Twitter, and many more companies, which represent, model, and process their data as graphs. If we look at the world wide web, that can also be viewed as a giant graph. This chapter demonstrates how real-world data science problems can be modeled as graphs.

## 5.1 Graph Definition

There are some representations of data that people most often misunderstand and treat as if they are graphs. One is the sales of some items against time, which gives a nice visual representation of the data, as shown in Figure 5.1. However, it is not a graph.

There is another very popular data representation, as shown in Figure 5.2, which is called a pie chart, and this also does not represent a graph.

Now, the question is why people often call these charts as graphs. This is because a chart usually depicts the graph of a function. Consider the data listed in Table 5.1 corresponding to the above pie chart shown in Figure 5.2. The first column represents the type of furniture available in

115

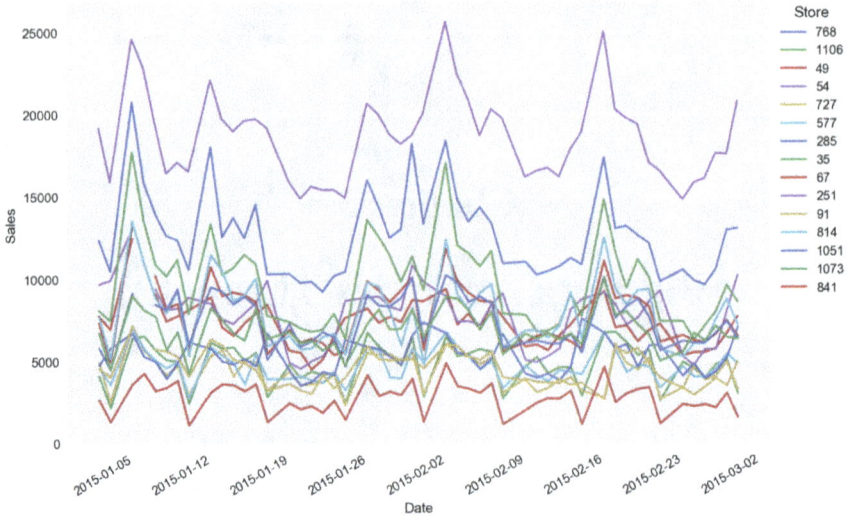

**Figure 5.1.**    Sales Data Representation [1]

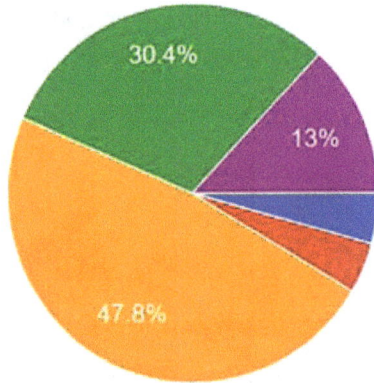

**Figure 5.2.**    Pie Chart

**Table 5.1.**    Data Corresponding to the Pie Chart

| Furniture Category | Percentage (%) |
| --- | --- |
| Office chairs | 47.8 |
| Almirahs | 30.4 |
| Shelves | 13 |
| … | … |

offices, and the second column represents the percentage of each category.

Now, we can define a mapping from the furniture category to percentage. The visual representation of this mapping can be represented in the form of a bar chart or a pie chart.

Graph analytics is founded on a branch of mathematics known as *graph theory*, and graph theory itself was born from a highly practical problem known as the Konigsberg bridge problem. Now, let's examine graphs from mathematical and computer science perspectives. From the perspective of mathematics, a graph is defined as a double set ($V$, $E$), where $V$ is the set of vertices and $E$ is the set of edges. From the computer science perspective as well, we need to adhere to the mathematical definition; however, here we need to represent and manipulate the information given in these graphs. For that, we need some form of data structure for its representation and operations for its manipulation. The data structures used to represent graphs are the adjacency matrix and the adjacency list.

### 5.1.1 Examples of Graph Analytics for Big Data

This section discusses different examples of graph analytics for big data in four disciplines, namely, social media, biology, human information network, and smart cities [2–4].

### 5.1.1.1 *Social Media*

In the first example of graph analytics for big data, we analyze Twitter and the graph depicting tweets, as shown in Figure 5.3 [5]. Tweets can include users, contain text, link to other tweets or URLs, have hashtags, and include references to a variety of media formats, and people publish tweets, respond to tweets, mention other users, etc. In Figure 5.3, the light-blue nodes represent tweets, whereas the purple nodes represent users. When a certain user creates a particular tweet, an object is generated in the graph.

Now, the question is: How does big data analytics comes into the picture? From the types of games played, their comments and discussions about various characters help behavioral psychologists predict if the people under study show violent behavior, if they are addicted to a game, etc. But why are graphs useful for this purpose? They are useful because,

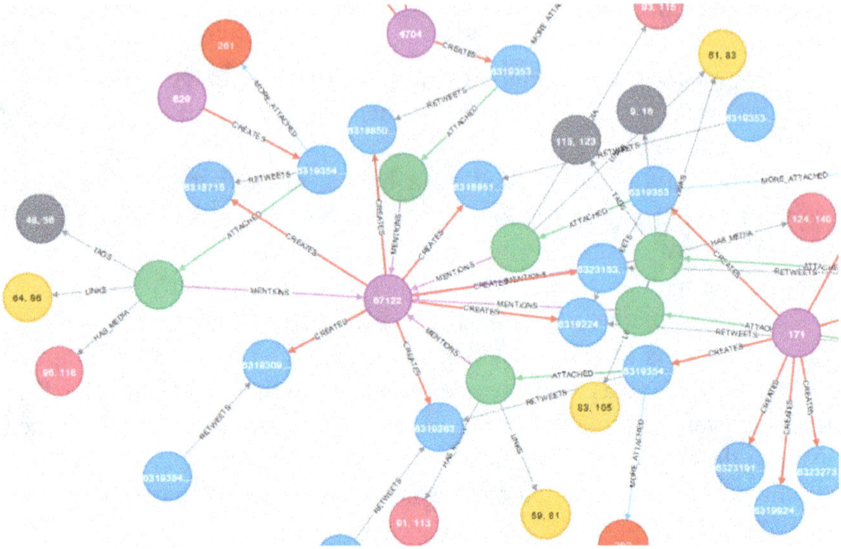

**Figure 5.3.**    Tweets in an Online Game

from the graphs, one can extract elements such as conversations. As far as we are aware, all users involved in this conversation are posting content, with someone reacting to the post and somebody else responding to the response. They are retweeting and reacting, and so forth. We can observe a succession of conversations. Although one does not know if it is violent, there is at least a vigorous discussion over something. This conversation provides insight into the community of the people interacting, such as whether they are part of a group or who are the most prominent users to whom everybody listens. Graph analytics can be used to answer all such questions about a live conservation happening on a social media stream.

### 5.1.1.2 *Biological Networks*

This second use case relates to biology, where interaction arises naturally, such as genes producing proteins and proteins regulating the functions of other proteins. These interactions may demonstrate the presence of pathogenic disorders. All of these relationships can be arranged into networks of graphs, where the nodes are biological entities and the edges reflect various types of molecular interactions and illness linkages.

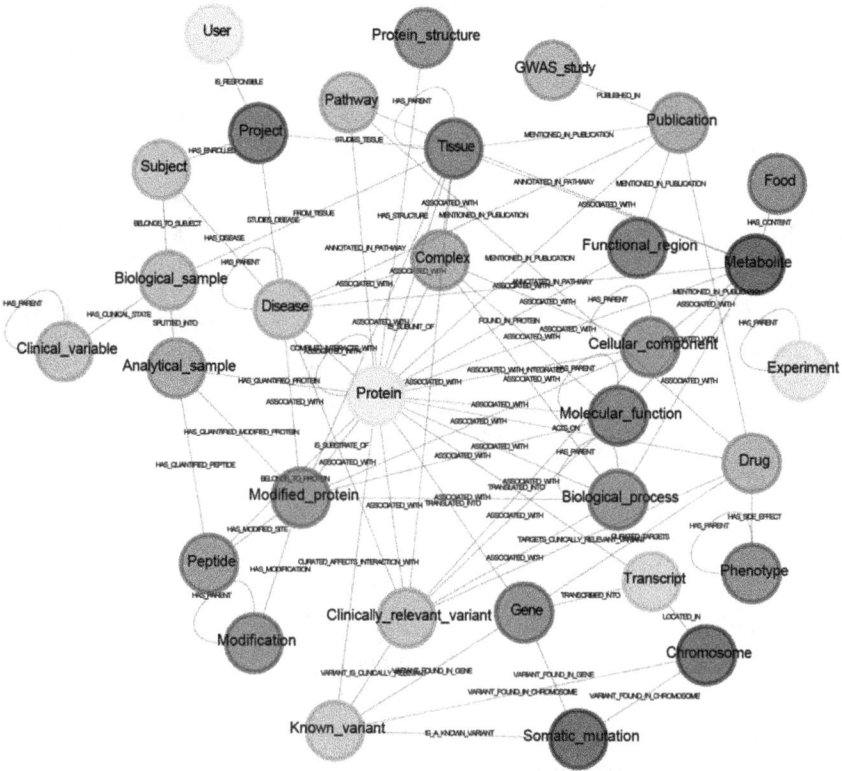

**Figure 5.4.**   Graph Resulting from Different Interactions Among Biological Entities [6]

Graphs like the one depicted in Figure 5.4 are typically compiled from a variety of data sources created by a variety of independent scientific groups with distinct research objectives and scientific methodologies, but with common biological entities. These networks may originate from experimental data, such as those pertaining to gene–protein correlations, gene–gene interactions, and gene-phenotype–disease connections. These may be the result of anatomical information about the human body or computational algorithms that mine the existing literature. These different logical components are stitched together to create such a graph, whose size grows as more results are linked and integrated into existing graphs, and this often leads to big data problems. As in the case of social media, here also we are interested in discovering unknown relationships. For instance, we can take two very different diseases, such as colorectal

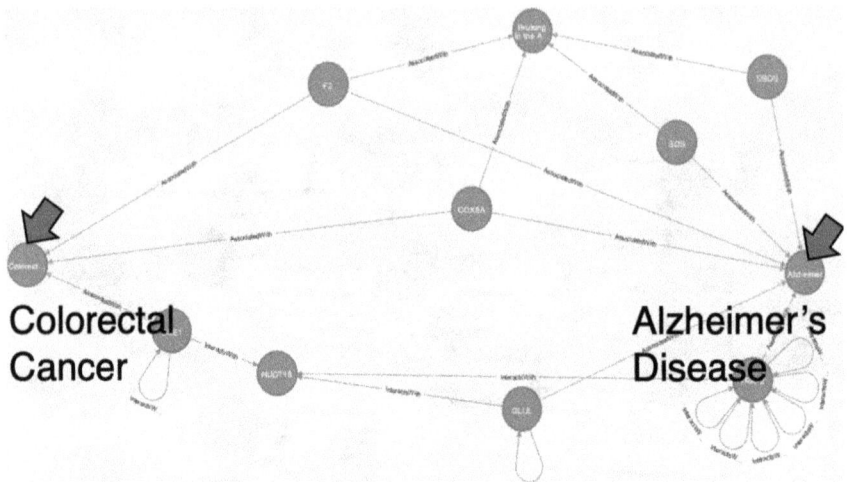

**Figure 5.5.**    Association Between Two Different Diseases [7]

cancer and Alzheimer's disease, and find out whether they are associated. If so, can we determine the underlying network linking them?

As can be seen in Figure 5.5, there are multiple genes that directly or indirectly link these two disorders. Thus, we can employ path-finding algorithms to uncover previously undisclosed network connections.

### 5.1.1.3 *Personal Information Networks*

Consider a personal information network, such as LinkedIn, where everyone that is on one's network is represented as a node, and if they know each other, this is represented with an edge, as shown in Figure 5.6.

It can be observed from the above figure that there are certain clusters/ groups that we should be able to identify automatically. With LinkedIn, we only have access to professional information. However, it is essential to inquire whether the professional network can be integrated with other types of personal data. This might include additional social networking data, such as the Facebook or Google friend network, or information such as the Outlook email network. By doing this, one can provide additional facts and interpersonal interactions. For example, Professor X, who is on

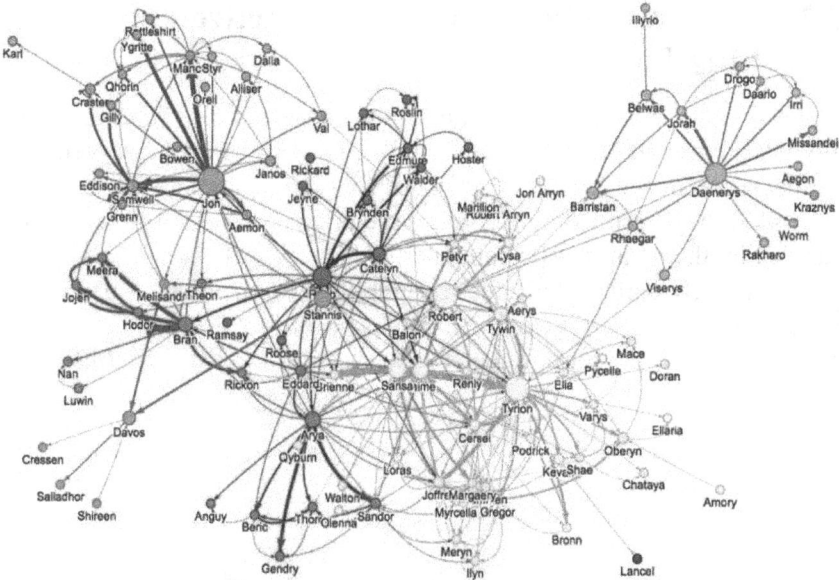

**Figure 5.6.** LinkedIn Personal Profile Network [8]

my contact list, is my director. Further, if one adds their schedule to this network, others can see the activities attended by that individual, etc. Taking it a step further, for specialized applications, one may integrate financial and business transactions, activity performance, or GPS location. This type of network or graph can be used for matching people with jobs; for example, to recruit for high-level positions, such as a company's board of directors, one must examine the candidate's network to determine their contacts with influential organizations, groups, and individuals. Choosing a surgeon based on his or her social media ratings would be another example. We may also wish to identify individuals who can impact a human network. Suppose we are on an election campaign team and need to reach as many individuals as possible in a city with our message. We cannot walk door-to-door ourselves; instead, we must reach out to certain individuals who will relay our message to others. Graph analytic tools may assist in identifying the smallest number of individuals who can contact the greatest number of potential voters in a city.

## 5.2  Graph Analytics from the Perspective of Big Data

When we talk about graph analytics in the context of big data, we generally talk about the impact of the four V's, namely, the volume, velocity, variety, and valence of graph data. If we consider a dataset, such as a load network, the corresponding graph is quite large compared to what can be handled by the memory of a computer.

Now, we examine the effect of graph size on analytical processes. What is the meaning of velocity in graphs? Consider the graph of Twitter discussion about the European elections, as shown in Figure 5.7 [9].

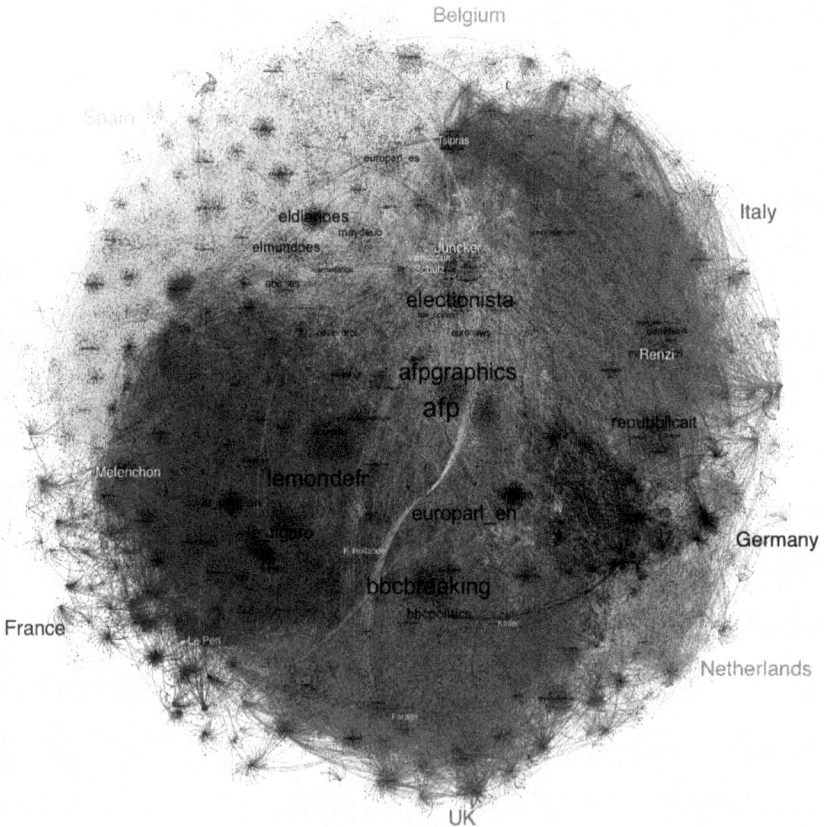

**Figure 5.7.**   Graph of Twitter Discussion [9]

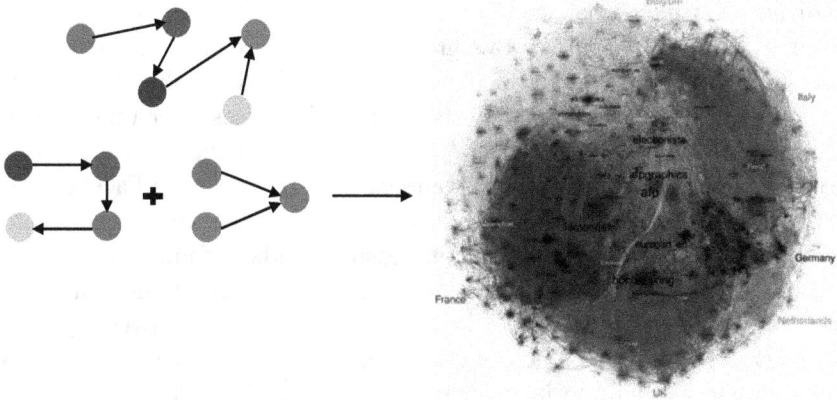

**Figure 5.8.**   Creation of Twitter Discussion Graph

When someone writes a post that will be liked and commented on by other users, it creates little graphs, as shown in Figure 5.8, and there can be numerous such updates that contribute to the resultant graph shown in Figure 5.7.

After some time, if a similar action is repeated, it is again added to the graph. Therefore, as time passes, more and more edges get added to the graph, and the rate at which this happens is really high, and it is referred to as the streaming of edges into graphs.

Further, there can exist multiple streams at the same time. This indicates that the graph is collecting data from multiple sources, and each of these distinct sources contributes distinct types of data to the graph, which is called *variety* in the context of graphs. There are two facets to variety. One, as we have already stated, graph data are frequently formed by integration [10], as we saw in the example stated above. Consequently, there is a great deal of variability because the nature of the data is diverse. For instance, the data may originate from a relational database or an XML database, a document, or complex entities, such as social networks, citation networks between articles or patents, interaction networks, and web entities connected by links.

The next V is valence. The concept of valence is derived from valence electrons, which are bonding electrons within an atom, and the remaining electrons are known as core electrons. This concept in the context of graphs tells us that as we raise the valence of the graphs, the connectedness of the graphs will grow. Apart from connectedness, valence also indicates the interdependence of data [11]. Therefore, if the valence is high, it indicates that there are a greater number of data components that are highly related, and these links can be utilized. In the majority of instances, valence becomes significant as it increases over time, which causes the network to get denser and the average distance between nodes to decrease. For example, when someone first begins using Facebook or any other social media, they will have a small number of users who are not actually related. Over time, an increasing number of individuals begin connecting, and with time, one can observe thick groupings inside the network since the knowledge and interconnectedness between people have evolved and grown denser over time. This is the valence phenomenon, which is crucial to investigate since one needs to determine which areas of the graph have become denser and why they have grown denser. Perhaps there was an event that brought these individuals together, and we wish to determine this using graph analytics.

## 5.3  Techniques for Graph Analytics

In this section, we discuss several graph analytics techniques and develop an understanding with which one can determine the right technique for the given graph analytics problem. This section covers graph analytics techniques with a greater focus on their mathematical and algorithmic aspects. There are four classes of graph analytics covered in this section, namely, path analytics, connectivity analytics, community analytics, and centrality analytics [12]. Path analytics is a type of analytic technique in which the primary purpose is to traverse a graph's nodes and edges. Connection analytics investigates the connectivity pattern of a graph, which refers to the organization and structure of the graph's edges. Community analytics involves tracing the activity of communities, which are entities in a network that interact closely with one another. Centrality analytics identifies and classifies significant network nodes in relation to a certain analysis problem.

### 5.3.1  Basic Definitions

We start with the basic mathematical definition of graphs. A graph is a double set of vertices/nodes and edges, where each element can be viewed as an ordered pair. However, we need to extend this definition to include

**Table 5.2.** Additional Information Elements in a Graph

| Extended Graph Definition | |
|---|---|
| Component | Description |
| $V$ | Set of vertices |
| $E$ | Set of edges |
| $NT$ | Set of node types |
| $f{:}NT \rightarrow V$ | Type assignment to nodes |
| $ET$ | Set of edge types |
| $g{:}NT \rightarrow E$ | Type assignment to edges |
| $NA$ | Set of node attributes |
| $EA$ | Set of edge attributes |
| $dom(NA[i])$ | Domain of the $i$th node attribute |
| $dom(EA[i])$ | Domain of the $i$th edge attribute |

other information elements, as shown in Table 5.2, since a real-life graph has more information content.

Consider the example of a Tweet graph, as shown in Figure 5.9.

A Tweet is a complex output of information because it is a network with multiple nodes and edges. There are multiple types of nodes in a Tweet. For instance, it contains Tweet nodes, user nodes, media nodes, URL nodes, and hashtag nodes. This process of assigning types or labels to nodes is sometimes referred to as *node typing*. Mathematically, we can add two extra elements to our initial definition of graphs. The collection of node types comes first, followed by the mapping function that decides which node type goes with which collection. Further, a node possesses attributes and values in addition to the different types. For a certain type of data, such as a Tweet, Twitter has determined a defined set of attributes. This set of attributes is referred to as a *node schema*. The number of attribute–value pairs in a node structure is completely up to the designer. Much like nodes in a network have nodes, edges have edge types, also called *edge labels*. Edges, like nodes, can have their own attribute–value pair-based schemas.

Numerous applications encode distinct types of quantitative information as edge weights in a graph. When there are no weights associated with an adjacency metric, an edge is represented by entering "1" into the

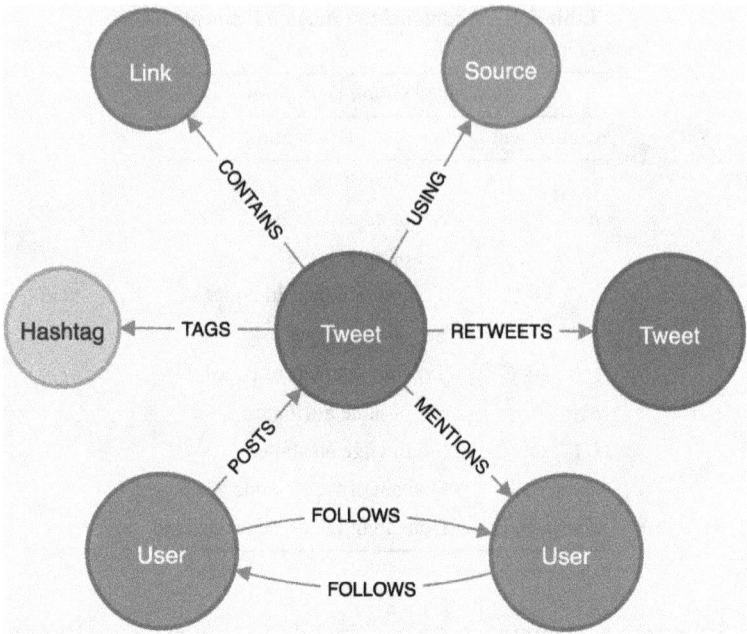

**Figure 5.9.**   Tweet Graph [13]

cell that corresponds to it. In any case, if we want to use a weight, the value of the weight can be incorporated into the adjacency matrix in order to facilitate computations later on down the line. What does the edge weight indicate? This is dependent on the application. Let's see some examples. The most obvious example is a map of roads, in which the nodes represent intersections of roads and the edges indicate sections of roads or highways that connect these nodes. A given road segment's distance can be represented by the edge weight of that section. The quality of the connection between a sender and a receiver can be inferred within the context of a personalized communication network, such as an email network, by analyzing the average number of messages sent each week between the two parties. So, more emails equal a better connection. Within a biological network, the chemical environment at the site of a reaction can be predicted with respect to the levels of the reactants provided, etc. It is often necessary to determine whether any possible interaction is likely to occur. This is represented as a weight that indicates the interaction probability. Finally, consider a knowledge network in which nodes represent entities such as individuals, locations, and events, and edges reflect relationships, such as some famous personality dating some person *Y*. Now, it's possible that some news organizations could find value in this kind of information. Therefore, it is prudent to assign a certain grade to the information, even if it does not originate from a credible source. This level of confidence can serve as a weight on the edge if it is applied appropriately.

A lot of times, essential information may be gleaned from the structure of a graph. A loop is an example of one of these structural properties. A loop is formed when an edge connects a node back to itself. There are several examples of loops in graphs, such as people writing emails to themselves and a web page that links to its own URL. Loops can also be seen in real-world situations, such as when a road segment returns to the same crossroads. The presence of loops and the nodes that contain them can be very enlightening in certain applications, yet in other contexts, they can be a source of potential complications. In addition, a graph has the structural quality of having numerous edges connecting the same pair of nodes, which is another one of its properties. The graphs with this characteristic are known as multi-graphs. This is a common occurrence in human networks. A person can simultaneously be someone's husband, co-performer in music, and financial advisor. Numerous analytics methods are not innately designed to deal with multi-graphs and frequently require some modification to handle them.

## 5.3.2 Path Analytics

This section describes what paths are, how they differ from walks, and why they are frequently more beneficial in real-world applications. Consider a graph where each edge has some weight associated with it, as shown in Figure 5.10. One way to think of this is as the graph of a road network, with the cities serving as the nodes and the edge weights indicating the distance that separates each pair of cities.

A walk is any chosen path from one node to another, connected by edges; for example, H→F→G→C→F→E→B is a walk in which we passed by node F twice. In many cases, we do not want to investigate arbitrary walks but rather a walk in which no nodes are repeated until we are required to return to the starting point. This is because arbitrary walks can lead to unexpected results. Such a restricted route is known as a path, as shown in Figure 5.11. The green arrows denote a path between J and B.

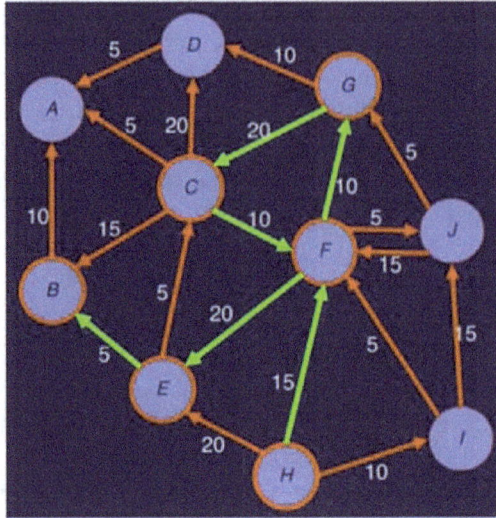

**Figure 5.10.** Random Walk in a Simple Weighted Graph

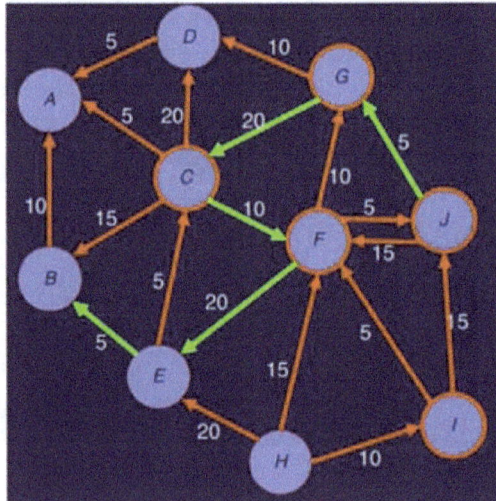

**Figure 5.11.** Path in a Simple Weighted Graph

In addition, if a path has three or more nodes and begins and termi-
nates on the same node, it is considered a cycle. A trail resembles a path
in which no edge can repeat itself. The next notion to consider is that of a
graph's diameter. If we start at any arbitrary node in a graph and go to any

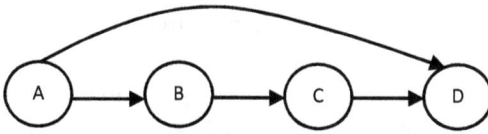

**Figure 5.12.**  Diameter of a Graph

| | A | B | C | D | E | F | G | H | I | J |
|---|---|---|---|---|---|---|---|---|---|---|
| A | 0 | ∞ | ∞ | ∞ | ∞ | ∞ | ∞ | ∞ | ∞ | ∞ |
| B | 1 | 0 | ∞ | ∞ | ∞ | ∞ | ∞ | ∞ | ∞ | ∞ |
| C | 1 | 1 | 0 | 1 | 2 | 1 | 2 | ∞ | ∞ | 2 |
| D | 1 | ∞ | ∞ | 0 | ∞ | ∞ | ∞ | ∞ | ∞ | ∞ |
| E | 2 | 1 | 1 | 2 | 0 | 2 | 3 | ∞ | ∞ | 3 |
| F | 3 | 3 | 2 | 2 | 1 | 0 | 1 | ∞ | ∞ | 1 |
| G | 2 | 2 | 1 | 1 | 3 | 2 | 0 | ∞ | ∞ | 3 |
| H | 3 | 2 | 2 | 3 | 1 | 1 | 2 | 0 | 1 | 2 |
| I | 4 | 3 | 3 | 3 | 2 | 1 | 2 | ∞ | 0 | 1 |
| J | 3 | 3 | 2 | 2 | 2 | 1 | 1 | ∞ | ∞ | 0 |

**Figure 5.13.**  Hop Distance Matrix

other arbitrary node in the graph by following only the shortest pathway routes, what is the greatest number of steps you are required to take? This maximum number is determined by the diameter of the graph. We can't say that diameter is the longest path between any two nodes in a graph. Consider the graph shown in Figure 5.12. Here, the diameter is not 3; in fact, it is 2.

This is because, here, we have a path of length 1 from A to D in addition to the path of length 3. Therefore, to find the diameter, we consider the longest path from B to D, which is of length 2.

Now, let's see how it is computed. Consider a matrix called the hop distance matrix, as shown in Figure 5.13, in which, as in an adjacency matrix, the nodes in the graph are represented by the rows and columns of the matrix, and each of its cell, $c_{ij}$, contains the distance from an $i$th node to a $j$th node via the shortest path. The distance of a node from itself is 0, and that of the node that is not reachable from the other is infinity.

From Figure 5.13, it can be observed that after filling all the entries, the largest reachable value is 4, which is the diameter of the given graph.

The question that pertains to path analytics that is the most fundamental is "what is the optimal or best path from one node to another?" [14]. In order to establish the optimal path, we must first determine the conditions under which one path is more advantageous than another. The majority of the time, this is formulated as an optimization problem, in which our subfunction must be minimized and maximized subject to constraints. There are generally two types of constraints: inclusion and exclusion criteria. Inclusion criteria can describe which nodes are required to be part of the path, while exclusion criteria can outline which nodes and edges should not be part of the path. In addition, one has the option of specifying a preference criterion that either acts as a more rigorous or a looser constraint. For example, we would like to minimize our use of highways or avoid congestion on my trip. Although the users would prefer that they be fully enforced, it is acceptable if they are not completely enforced. My commute to work in the morning is a good example of a practical use case. I would prefer to travel the shortest distance possible from my home to my office. However, I must drop off my son at school. Therefore, my path must include his school, which is one of my preferences. Another preference criterion is that I would like to avoid the roads surrounding the new construction because there is usually a significant amount of traffic congestion there. Thus, we can incorporate the various preference criteria. This issue is addressed by all mapping and navigation software. Here is a screenshot of Google Maps depicting my journey from my current location at NITJ to a model town, Jalandhar, as shown in Figure 5.14.

Google Maps displays three distinct routes and highlights the best option. The actual shortest path of 12 km will take the longest amount of time at the time of day when I referred to the map. Therefore, the weights listed here are not distances but rather estimated travel times. We should also note that Google presents the blue, red, and orange segments of the preferred path. The orange and red street segments clearly represent congested areas and have a greater weight than the blue segments as a result. Therefore, the weights of the street segments are not truly static, but vary based on numerous other variables, such as the weather and the time of day. For this reason, it is essential to find a solution to the problem relating to the way with the least amount of weight in order to help commuters.

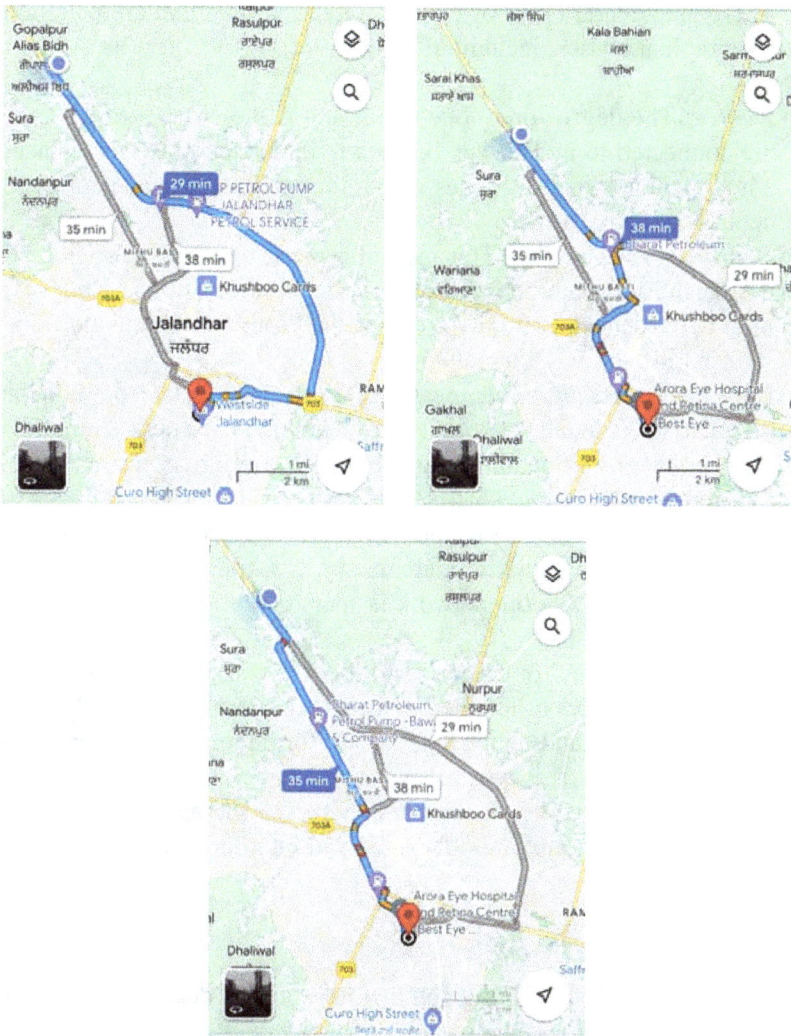

**Figure 5.14.**   Google Map Snapshot

## 5.3.3 Connectivity Analytics

Connectivity analytics is a type of analysis that is commonly used in graph analytics. In graph theory, connectivity refers to the relationship between the nodes in a graph, and it can be used to understand the structure and behavior of a network [15].

There are several types of connectivity measures that are commonly used in graph analytics, including:

- *Degree*: The degree of a node in a graph is the number of edges that are connected to it. This can be used to understand the importance of a node in a network, as nodes with higher degrees are generally more central to the network.
- *Betweenness centrality*: This measures the extent to which a node lies on the shortest path between other nodes in the network. Nodes with high betweenness centrality are important for maintaining the flow of information through the network.
- *Closeness centrality*: This measures the distance between a node and all other nodes in the network. Nodes with high closeness centrality are generally more influential and have more control over the network.
- *Clustering coefficient*: This measures the extent to which nodes in a network tend to cluster together. High clustering coefficients indicate that the network is highly interconnected, while low clustering coefficients indicate that the network is sparser.

By using these and other connectivity measures, graph analytics can help identify key nodes in a network, detect communities and clusters within the network, and understand how information and influence flow through the network. This type of analysis is particularly useful for understanding social networks, transportation networks, and other complex systems where nodes are interconnected and interdependent.

### 5.3.4 Community Analytics

Communities are clusters in a graph that have a high degree of interaction. Community analytics, also known as community detection or clustering, is a type of analysis commonly used in graph analytics. It involves identifying groups of nodes, known as communities or clusters, that are more densely connected to each other than to nodes in other parts of the graph [16]. This can be used to identify groups of related or similar nodes within a larger network and to understand the structure and behavior of the network as a whole.

There are several methods used for community detection in graph analytics, including:

- *Modularity-based methods*: These methods seek to maximize the modularity of the network, which is a measure of the density of connections within communities compared to connections between communities.
- *Spectral clustering*: This method involves representing the network as a matrix and using linear algebra to identify groups of nodes that are more closely connected to each other.
- *Hierarchical clustering*: This method involves iteratively merging smaller clusters into larger ones based on their similarity or distance from each other.
- *Random walk-based methods*: These methods involve simulating random walks through the network and using the resulting paths to identify communities.

By using these and other methods, community analytics can help identify subgroups of nodes within a larger network, understand the relationships and interactions between these groups, and identify key nodes or communities that are particularly influential or important within the network. This type of analysis is particularly useful for understanding social networks, biological networks, and other complex systems where nodes are interconnected and interdependent.

The questions about community analytics are divided into three categories: static analysis, temporal/evolution analysis, and predictive analysis.

When doing static analyses, the queries pertaining to the analytics do not depend on the passage of time. In this section, we enquire about the makeup of the community, namely who is a part of the community and the degree to which its members are connected to one another. In the second category, we investigate the origins of the community as well as its historical development. Communities can arise from the occurrence of an event, such as an annual cultural festival; such types of communities are transient in nature. Here, we want to know how and why did such communities form or dissolve. Some communities, such as a political party or a group of Facebook members, are expected to persist throughout time and are therefore not considered to be transient. The third category is concerned with making forecasts; in this area, the analysts aim to forecast how a community develops through time and whether or not an individual will continue to be a part of this community.

Before we can ask questions about these communities, the initial step is to locate clusters of nodes inside the network. The idea that there are a greater number of connections inside a community than there are among communities needs to be standardized in order to truly be capable of recognizing communities. The degree of a node can be broken down into its internal and external components, which is one way of achieving this goal. The number of community edges represents the internal component of the model. The number of edges that are located outside of the community is represented by the exterior degree. The next thing that needs to be done is to take both the internal and external degrees of a cluster into account as a whole. When all of the nodes in a cluster have an internal degree, those degrees may be summed up together to get the overall internal degree of the cluster. In a similar manner, the external degree of the cluster can be determined by adding up the external degrees of each of the nodes. We can now define intra-cluster density as the percentage of edges within a cluster that are connected to one another:

$$\text{Intra-cluster density}(\delta_{\text{int}}) = \frac{\text{Number of internal edges in } c}{n_c(n_c - 1)/2}$$

where $C_2^{n_c}$ represents the number of possible internal connections and $n_c$ is the number of nodes in cluster $c$ [15]. Similarly, the inter-cluster density is the ratio of the number of edges between clusters to the number of possible pairings between cluster nodes $n_c$ and nodes $(n - n_c)$ that are not part of the cluster [13]:

$$\text{Inter-cluster density}(\delta_{\text{ext}}) = \frac{\text{Number of inter-cluster edges of } c}{n_c(n - n_c)}$$

There are two categories of network community discovery techniques. One of them places emphasis on local properties, which are properties that are investigated for a node in conjunction with its neighbors. A subgraph in which every node in the subgraph is connected to every other node in the subgraph constitutes the ideal community for a network. *Clique* is a term for this kind of social grouping. Finding the greatest clique in a graph — the optimal community organization as a clique — is a computationally hard job. Knowing the value of $k$, where $k$ is the number of individuals in a clique, makes it much simpler to identify and locate them. In the real world, it is more difficult to find cliques with more than three or four members. Therefore, we must relax its definition. There are now two categories of relaxation: distance-based and density-based. $n$-clique and $n$-clan are common topics of research in distance-based relaxing. An $n$-clique is a subgraph in which every pair of nodes has a distance of $n$ or less from one another. Connecting nodes between $n$-clique members are omitted. $n$-clan is a solution to this issue; in an $n$-clan, the shortest path between any two members, without involving outsiders, must be $n$. Finding coherent clusters of communities can be done with distance-based metrics, such as $n$-clique and $n$-clan, or with density-based methods, such as $k$-core. Each node in a $k$-core graph has a connection to exactly $k$ additional nodes.

The other category of community discovery focuses on global properties. The specific global property that we focus on is modularity. It is a global measure of a cluster quality [17]. If we examine the edges in a group and compare them to what would be observed if the edges were assigned randomly according to a probability distribution, we may find that they are distinct. If a community exists, there will be more edges than would occur by chance. If there is no community in a section of the graph, the number of edges in that section will be close to or even lower than the random case. Modularity measures the deviation between the true and false edge fractions in a graph to provide an evaluation of cluster quality. Modularity quantifies the number of edges that exist within a set minus the number of edges that would exist if they were randomly distributed. Mathematically, cluster quality can be formulated as

$$Q = \frac{1}{2m} \sum_{ij} \left( A_{ij} - P_{ij} \right) \cdot \delta \left( C_i - C_j \right)$$

where, $m$ is the number of edges, $A$ is the adjacency matrix, $P$ is the expected value of the probability of nodes $i, j$ to be connected under some probability model, $C$ denotes clusters, and $\delta$ is 1, if $i, j$ are on the same cluster, otherwise it is 0.

The adjacency matrix gives us the actual edges, $P_{ij}$ provides the probability of a random edge, and the delta function's task is to evaluate if $i$ and $j$ should be in the same cluster. If they are, the contribution will be added to $Q$, the quality metric. There are many ways to look at what the probability model $P_{ij}$ looks like. According to a simple model, the probability of an edge between nodes $i$ and $j$ is proportional to the degree of node $i$ multiplied by the degree of node $j$. So, if nodes $i$ and $j$ are already well interconnected, there is a significant likelihood that they share an edge.

### 5.3.5  Centrality Analytics

Centrality analytics is an important concept in graph analytics, which is the process of analyzing and interpreting data represented in the form of a graph. In graph analytics, centrality measures the importance of nodes or edges within a graph.

There are several types of centrality measures, including degree centrality, betweenness centrality, closeness centrality, and eigenvector centrality:

- Degree centrality measures the number of edges that are connected to a node, and it is used to determine the popularity of a node. Nodes with a high degree of centrality are considered to be highly connected and, therefore, important.
- Betweenness centrality measures the number of times a node acts as a bridge between other nodes in the graph. Nodes with high betweenness centrality are important for maintaining the connectivity of the graph and facilitating the flow of information between nodes.
- Closeness centrality measures how close a node is to all the other nodes in the graph. Nodes with high closeness centrality are important for quickly disseminating information across the graph.
- Eigenvector centrality measures the influence of a node based on the influence of its neighbors. Nodes with high eigenvector centrality are considered to be highly influential within the graph.

Centrality analytics is important in a variety of fields, including social network analysis, transportation analysis, and biological network analysis. It can be used to identify key players in a network, analyze the flow of information or goods through a network, and understand the dynamics of biological processes.

## 5.4  Large-Scale Graph Processing

In this section, we discuss the fundamental principles of large-scale graph processing and the supporting software architecture. Here, we examine the concept of the programming model and introduce the bulk synchronous parallel (BSP) programming model, which is created specifically for graph-oriented processing. There are two well-known implementations of this concept: Google's Pregel [18] and Carnegie Mellon University's Graphlab [19].

## 5.4.1 Parallel Programming Model for Graphs

Big data computation is based on parallel computation [20]. Obviously, for a program to be parallel, a number of concurrently functioning processes are required. But how do these processes exchange information and communicate? How do they determine the timing of their interactions? Moreover, what precisely occurs in parallel? Consider the initial question. Two processes share memory to exchange data. There are architectures that can make the memory of several machines appear as though it were one large addressable memory area. Yet, two processes will also communicate via the exchange of messages, either directly from one process to another or via a common message-carrying conduit, sometimes known as a message bus. The second question has various answer options as well. Task parallelism and data parallelism are two of the most frequent approaches to achieving parallelism. In task parallelism, a big task can be subdivided into numerous concurrently executable subtasks. In data parallelism, data can be partitioned into numerous smaller chunks, and operations can perform independently on each partition. Generally, these partial operations and partially processed data are synchronized and concatenated to form a complete response. We must keep in mind that task parallelism is slightly distinct from data parallelism, and it is conceivable for a programming model to contain both types of parallelism.

Further, we should note that a programming language is independent of the programming model. Thus, a programming model can be implemented in numerous languages. As previously said, the programming model we will investigate is BSP. BSP was not initially designed for graph processing. It was conceived as a parallel computing model to bridge the gap between software models of parallel processing and hardware support for parallelism. The BSP concept is as follows.

In BSP, there are a number of processors, and each processor is capable of performing computations locally, utilizing its own local memory. There also exists a router, which can forward a message from any processor to any other processor. When two nodes exchange messages, a third node can continue to do computation. BSP provides a facility by which the state of all or a subset of processes can be synchronized. This synchronization may occur periodically, at $T$-second intervals, or another method may be used to indicate when it will occur. But when it happens, all affected processors will reach a consistent state. When synchronization is performed, a new computation cycle can begin. This is referred to as the point of synchronization or barrier synchronization because all processes must reach this barrier or threshold before proceeding to the next processing step. A BSP program is composed of a series of super steps. If necessary, each CPU will get the data during each super step and, If necessary, perform computation and then share the data with the appropriate partner. Once all nodes are complete, the system synchronizes, and afterward, the subsequent round begins. Each processor can determine whether it must perform computations or data exchanges. Otherwise, it will become inactive. If necessary, a processor can be reactivated at a later time. When all processors are idle, computation stops.

In applying the BSP model to graphs, several assumptions are made. Each processor is assumed to be synonymous with a node in the graph, and a processor can only send messages to or receive messages from its neighboring processes. We also assume that vertices have IDs and perhaps complex values, and further, the edges also have IDs. Each vertex is aware of the edges to which it is related. A computation can now be viewed as a vertex-centered activity. This programming paradigm is known as "think like a vertex". In order to think like a vertex, we must understand its capabilities. A vertex is capable of identifying itself, get/set its value, get/count its edges, get/set a specific edge's value using the edge ID or the ID of the target vertex, get values of all edges connected to a vertex, and add/remove a specific edge. Lastly, since vertices are processes, they can initiate or terminate computation. Nodes typically awaken when they receive a message from another node. Compared to a vertex, an edge has significantly fewer capabilities. It can just get its own ID if the system allows edge IDs, it can set and retrieve its own values, and it can get the ID of the node it is pointing to.

## 5.5 Chapter Summary

In this chapter, we have learned about graph analytics, which is the process of analyzing and interpreting data stored in graph structures.

Graphs are used to represent data as nodes and edges, where nodes represent entities such as people, places, or things, and edges represent the relationships between these entities. When it comes to big data, graph analytics can be a powerful tool for discovering patterns and insights that may be difficult to uncover using traditional data analysis methods. The following are some key considerations for performing graph analytics on big data: (i) Choose the right graph database: When dealing with large volumes of data, it's important to choose a graph database that can handle the scale and complexity of the data. Some popular graph databases for big data include Apache Cassandra, Neo4j, and Amazon Neptune. (ii) Use distributed computing: To perform graph analytics on big data, it's often necessary to use distributed computing frameworks, such as Apache Spark or Hadoop. These frameworks allow for parallel processing of data across multiple nodes in a cluster, enabling faster processing of large volumes of data. (iii) Choose the right algorithms: There are a wide variety of graph algorithms that can be used for different types of analysis. Some popular algorithms for graph analytics on big data include PageRank, community detection, and centrality measures, such as betweenness and closeness. (iv) Visualize the results: Graph visualization can be a powerful way to understand the results of graph analytics on big data. Tools such as Gephi and D3.js can be used to create interactive visualizations of graph data that can help uncover patterns and insights.

# References

1. Pavlyshenko, B. M. (2019). Machine-learning models for sales time series forecasting. *Data*, 4(1), 15.
2. Nie, X., Peng, J., Wu, Y., Gupta, B. B., & Abd El-Latif, A. A. (2022). Real-time traffic speed estimation for smart cities with spatial temporal data: A gated graph attention network approach. *Big Data Research*, 28, 100313.
3. Psannis, K. E., Stergiou, C., & Gupta, B. B. (2018). Advanced media-based smart big data on intelligent cloud systems. *IEEE Transactions on Sustainable Computing*, 4(1), 77–87.
4. Cuzzocrea, A. & Song, I. Y. (2014). Big graph analytics: the state of the art and future research agenda. In *Proceedings of the 17th International Workshop on Data Warehousing and OLAP*, 3–7 November 2014, Shanghai, China, (pp. 99–101).
5. Wallner, G., Kriglstein, S., & Drachen, A. (2019, August). Tweeting your destiny: Profiling users in the twitter landscape around an online game.

In *2019 IEEE Conference on Games (CoG)*, 20–23 August 2020, London, United Kingdom, (pp. 1–8). IEEE.

6. Santos, A., Colaço, A. R., Nielsen, A. B., Niu, L., Geyer, P. E., Coscia, F., Albrechtsen, N. J. W., Mundt, F., Jensen, L. J. & Mann, M. (2020). Clinical knowledge graph integrates proteomics data into clinical decision-making. *bioRxiv*, https://doi.org/10.1101/2020.05.09.084897.

7. Ospina-Romero, M., Glymour, M. M., Hayes-Larson, E., Mayeda, E. R., Graff, R. E., Brenowitz, W. D., Ackley, S. F., Witte, J. S., & Kobayashi, L. C. (2020). Association between Alzheimer disease and cancer with evaluation of study biases: A systematic review and meta-analysis. *JAMA Network Open*, 3(11), e2025515–e2025515.

8. Anand, R. (2020). An illustrated guide to graph neural networks. *Medium, Democratizing Artificial Intelligence Research, Education, Technologies (Dair.ai)*, 30, https://medium.com/dair-ai/an-illustrated-guide-to-graph-neural-networks-d5564a551783.

9. Smyrnaios, N. (2014). EP2014: The European election through the lens of Twitter, http://ephemeron.eu/1219.

10. Psannis, K. E., Stergiou, C., & Gupta, B. B. (2018). Advanced media-based smart big data on intelligent cloud systems. *IEEE Transactions on Sustainable Computing*, 4(1), 77–87.

11. Vinoth, R., Deborah, L. J., Vijayakumar, P., & Gupta, B. B. (2022). An anonymous pre-authentication and post-authentication scheme assisted by cloud for medical IoT environments. *IEEE Transactions on Network Science and Engineering*, 9(5), 3633–3642.

12. Nisar, M. U., Fard, A., & Miller, J. A. (2013). Techniques for graph analytics on big data. In *2013 IEEE International Congress on Big Data*, 6–9 October 2013, Santa Clara, CA, USA, (pp. 255–262). IEEE.

13. "Graph Your Twitter Activity in Neo4j", http://network.graphdemos.com/.

14. Fu, L. & Deng, J. (2013). Graph calculus: Scalable shortest path analytics for large social graphs through core net. In *2013 IEEE/WIC/ACM International Joint Conferences on Web Intelligence (WI) and Intelligent Agent Technologies (IAT)*, 17–20 November 2013, Atlanta, Georgia, (Vol. 1, pp. 417–424). IEEE.

15. El Mouden, Z. A., Taj, R. M., Jakimi, A., & Hajar, M. (2020). Towards using graph analytics for tracking COVID-19. *Procedia Computer Science*, 177, 204–211.

16. Negara, E. S. & Andryani, R. (2018). A review on overlapping and non-overlapping community detection algorithms for social network analytics. *Far East Journal of Electronics and Communications*, 18(1), 1–27.

17. Gupta, S., Aga, D., Pruden, A., Zhang, L., & Vikesland, P. (2021). Data analytics for environmental science and engineering research. *Environmental Science & Technology*, 55(16), 10895–10907.

18. Malewicz, G., Austern, M. H., Bik, A. J., Dehnert, J. C., Horn, I., Leiser, N., & Czajkowski, G. (2010). Pregel: A system for large-scale graph processing. In *Proceedings of the 2010 ACM SIGMOD International Conference on Management of data*, 6–10 June 2010, Indianapolis, Indiana, USA, (pp. 135–146).
19. Han, M., Daudjee, K., Ammar, K., Özsu, M. T., Wang, X., & Jin, T. (2014). An experimental comparison of pregel-like graph processing systems. *Proceedings of the VLDB Endowment*, 7(12), 1047–1058.
20. Gaurav, A. & Santaniello, D. (2022). The development in data science due to the integration of blockchain technology. *Data Science Insights Magazine*, (Vol. 3, pp. 5–8). Insights2Techinfo.

# Chapter 6

# Taming Big Data with Spark 2.0

Apache Spark is an open-source big data processing framework that was designed to address the challenges of working with large datasets. It is one of the most popular big data processing tools, particularly for machine learning and data analytics. One of the key benefits of Spark is its ability to handle large-scale data processing tasks with speed and efficiency. Spark achieves this by distributing data processing tasks across multiple nodes in a cluster, allowing computations to be performed in parallel. Spark's core abstraction is the resilient distributed dataset (RDD), which is a fault-tolerant collection of elements that can be processed in parallel. Another advantage of Spark is its support for a wide range of data sources, including Hadoop Distributed File System (HDFS), Apache Cassandra, and Apache Kafka. This means that Spark can be used to process data from a variety of sources, making it a flexible tool for big data processing. In addition to its core processing capabilities, Spark includes several libraries and application programming interfaces (APIs) for machine learning, graph processing, and streaming data processing. These libraries, such as MLlib, GraphX, and Spark Streaming, make it easier for developers to build and deploy complex big data applications.

## 6.1 Introduction to Spark 2.0

The big data analytics framework Apache Spark [1,2] was created in 2012 at the University of California, Berkeley. Since then, it has drawn a lot of interest from both academics and businesses. It offers high-level APIs for Java, Scala, Python, and R, as well as an engine that is efficient and supports all execution graphs. It also supports a wide range of advanced tools, such as Spark Streaming, MLlib for machine learning, GraphX [3] for processing graphs, and Spark SQL [4] for processing SQL and structured data. It is a framework for broad data analytics on Hadoop-style distributed computing clusters, offering in-memory calculations to speed up and improve data processing over MapReduce. In addition to processing structured data in Hive and streaming data from HDFS, Flume, Kafka, and Twitter, Spark may run on top of an existing Hadoop cluster and utilize the Hadoop data store (HDFS).

## 6.1.1 Why Spark 2.0 Replaced Hadoop

Spark technology quickly became a viable alternative to the Hadoop MapReduce paradigm. Large dataset batch processing on massive commodity clusters is made simpler by MapReduce. Nevertheless, it has trouble handling iterative applications, such as machine learning and graph processing, which demand a lot of iterations. These two types of applications have relatively low MapReduce abstraction performance. Moreover, MapReduce is excessively sluggish and expensive for interactive applications. Take an input Wikipedia data file that has been split up by the MapReduce algorithm. Different workers are allocated to each division. The map function is then applied by each worker, producing intermediate results. As seen in Figure 6.1, "john, 1" will be generated

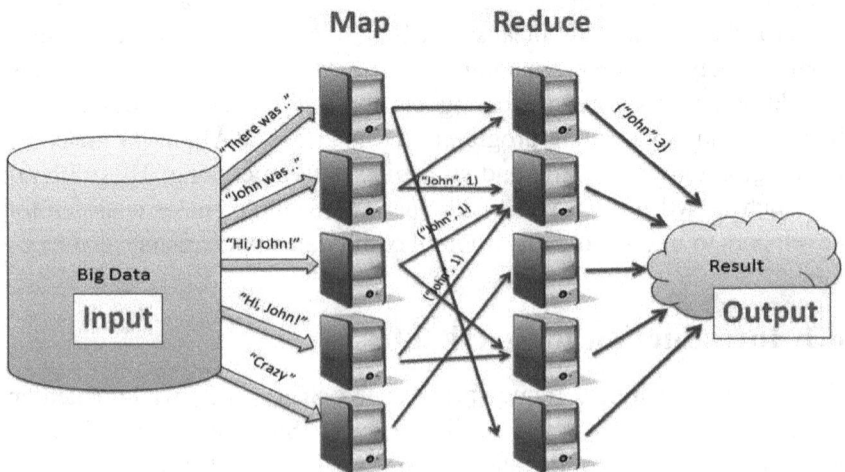

**Figure 6.1.**    Phases of Map Reduce on a Wikipedia Page

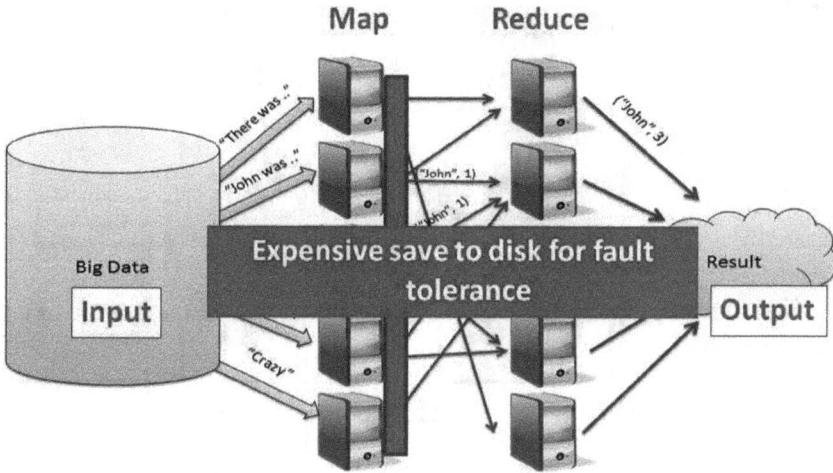

**Figure 6.2.**   Expensive Operations of Disks

during the map phase and saved in the HDFS, requiring numerous I/O operations. The storage of intermediate outputs on the file system, which requires several I/O transfers and slows down the overall computing process, is the main downside [5].

Even more overhead is created during the reduction step when the intermediate "key, value" pairs are read from the HDFS. Figure 6.2 illustrates this. The benefit of storing is that fault tolerance is achieved by making the intermediate data more persistent and enhancing durability.

Although fault tolerance can be accomplished between the map and reduce functions without any data loss, it comes at a high cost. The majority of real-world applications require multiple iterations of MapReduce, which slows the program down.

## 6.2  Resilient Distributed Datasets

Partitioned immutable datasets called RDDs [2] are collections of records, as shown in Figure 6.3. They can be cached for effective reuse and are constructed using coarse-grained transformations, such as map and join. A certain RDD might be cached or running in memory. As a result, it can address the issue of keeping intermediate results in the file system, which was noted as a weakness in the case of the Hadoop architecture.

Let's examine why Spark is necessary. The input file is divided when it enters the Spark framework and read into RDDs that operate on various

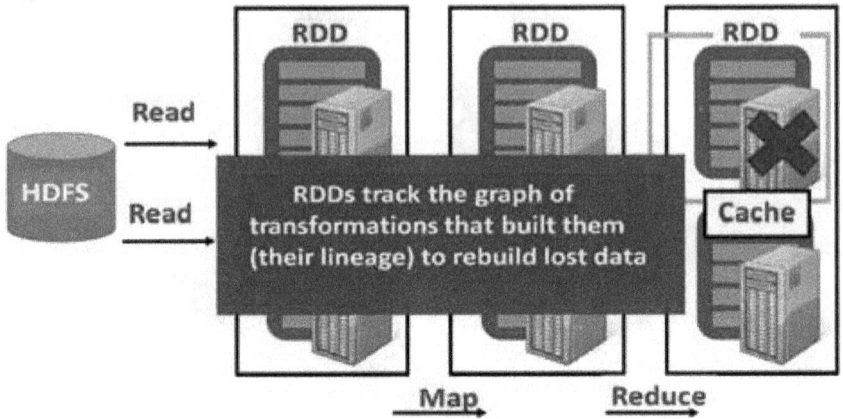

**Figure 6.3.**   Resilient Distributed Datasets (RDDs)

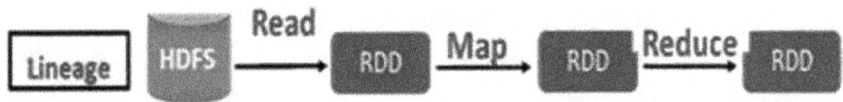

**Figure 6.4.**   Lineage in Spark

clusters. Then, the map function is used to transform and store the transformed RDDs in cache memory, eliminating the need to store the map function's intermediate outputs in the file system. When access to the intermediate file is required, the file system is employed. In-memory reads are a useful technique that can effectively increase performance by a factor of 100. The in-memory operations using RDDs are particularly effective because I/O operations can reduce performance. The cache will continue to hold the MapReduce operation's output. So, if more requests for the result are made through the cache, they can be made effectively. As a result, the Spark system can solve computations over the clusters with a faster compute speed thanks to this approach.

RDDs guarantee fault tolerance as well. RDD computations using map and reduce functions in memory are something we are already familiar with. Failures might be possible, but they might also be overcome through a recovery mechanism called *lineage*. This is depicted in Figure 6.4. It won't perform check pointing in order to support the lineage, but it will keep track of the logs of the coarse-grained operations carried out on the partitioned datasets.

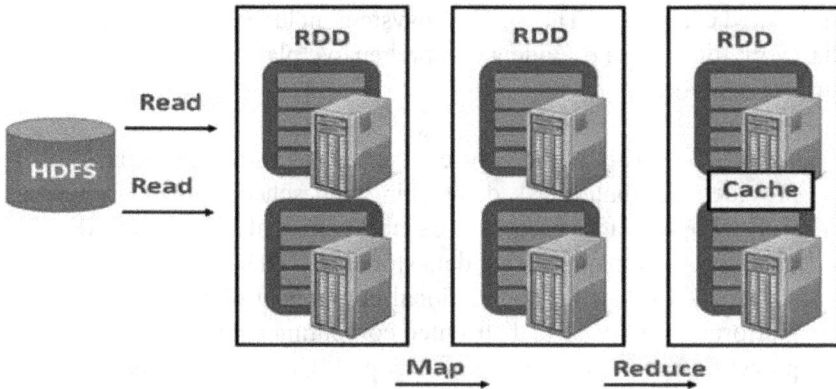

**Figure 6.5.**   Tracking of Graph Transformations in RDD

As a result, it will just recompute the missing partitions in the event of a failure. Conversely, if there is no failure, there will be no overhead cost associated with checkpointing.

It is established that the Spark framework, which uses lineage to reconstruct the RDD, can recover from the failure in addition to preserving the speed factor.

Consider a scenario where we must determine whether a specific RDD has crashed. If the previous stage failed, the Spark framework reconstructs the data contained in the RDDs utilizing the map phase through the log, performing specified operations on some RDDs in accordance with the log, while other RDDs are unaffected. In the event of failures, there are many coarse-grained procedures sufficient to recreate the RDDs.

When it comes to the lineage process, RDDs are used to read the data. To create the final RDDs, the map function is used, followed by a reduce operation. So, if there is a failure at any point, we can recreate the RDDs starting from that stage.

To reconstruct the deleted data, RDDs follow the graph of the transformations that formed them and their lineage, as seen in Figure 6.5. As a result, Spark's lineage feature supports the fault tolerance function.

## 6.3  Spark 2.0

Apache Spark is a powerful and popular open-source distributed computing system that is designed to process and analyze large datasets using distributed computing. The Spark ecosystem includes several components that work together to provide a comprehensive platform for data processing and analytics:

- *Spark Core*: This is the core component of the Spark ecosystem. It provides distributed task dispatching and scheduling capabilities as well as basic data processing capabilities. It also provides APIs for managing and manipulating data stored in RDDs.

  Spark Core is the foundational component of the Apache Spark platform, providing the distributed computing engine that enables the processing of large-scale datasets. Spark Core supports parallel processing of data using RDDs, which are an immutable distributed collection of objects that can be processed in parallel across a cluster of machines. Programming in Spark using RDDs involves creating RDDs, performing transformations on them, and then applying actions to obtain the results. Transformations create new RDDs from existing ones by applying operations such as filtering, mapping, and reducing. Actions are operations that return a value or write data to an external system, such as printing output to the console or writing data to a file.

  Spark pipelines are a way of chaining together multiple operations on RDDs to create a data processing workflow. Pipelines are composed of multiple stages, where each stage represents a set of transformations and actions that can be executed in parallel across the cluster. The key steps involved in programming Spark using RDDs in pipelines are:
  - **Creating an RDD:** This involves loading data into an RDD using methods such as textFile() or parallelize().
  - **Transforming an RDD:** This involves applying transformations, such as filter(), map(), and reduce(), to create new RDDs that are derived from the original RDD.
  - **Chaining transformations:** This involves chaining multiple transformations together to create a pipeline of operations.
  - **Performing actions:** This involves applying actions, such as count() or collect(), to trigger the execution of the pipeline and obtain the results.
- *SparkSQL*: SparkSQL is a component of the Apache Spark big data processing engine that enables the processing of structured and semi-structured data using SQL-like syntax. This component provides a

high-level API for working with structured and semi-structured data in Spark. It allows you to run SQL queries on Spark data, including data stored in external databases and data streams.

It provides a programming interface to work with structured data using SQL queries, making it easier to integrate SQL queries with Spark programs. SparkSQL provides a high-performance, distributed SQL engine that can query data stored in various data sources, including Apache Hive, HDFS, and Apache Cassandra. SparkSQL supports various data formats, such as JSON, Parquet, and Avro, and also provides various data sources and connectors for relational databases, NoSQL databases, and cloud storage systems. It also supports a range of SQL operations, such as SELECT, FROM, JOIN, GROUP BY, and ORDER BY, allowing developers and data analysts to work with data in a familiar way. In addition to SQL, SparkSQL also supports the use of programming languages, such as Python, R, and Scala, making it a versatile tool for working with big data. SparkSQL also includes features for data exploration, data visualization, and data cleaning, making it a powerful tool for data analysis and processing.

SparkSQL provides a powerful way to explore and analyze large datasets. Here are some of the most common operations:

○   **Select:** The SELECT operation is used to choose specific columns from a table.
○   **Filter:** The FILTER operation is used to select rows that meet a certain condition.
○   **Group By:** The GROUP BY operation is used to group rows based on a particular column or set of columns and then apply an aggregate function to each group.
○   **Order By:** The ORDER BY operation is used to sort the rows in a table based on one or more columns.
○   **Join:** The JOIN operation is used to combine two or more tables based on a common column.
○   **Union:** The UNION operation is used to combine two or more tables with the same structure.
○   **Distinct:** The DISTINCT operation is used to remove duplicate rows from a table.
○   **Aggregation:** Aggregation functions, such as SUM, AVG, MAX, MIN, and COUNT, can be applied to a column or set of columns in a table.

- ○ **Window:** Window functions allow you to perform calculations across rows that are related to the current row.
- ○ **Subqueries:** Subqueries can be used to perform complex calculations and filter data based on the results of another query.

    These operations can be used in various combinations to perform complex data analysis tasks, such as data cleaning, transformation, and modeling. SparkSQL also supports the use of machine learning algorithms, allowing you to build predictive models on large datasets.

- *Spark Streaming*: Spark Streaming is a component of Apache Spark that allows you to process real-time streaming data. It provides a high-level API for processing data streams, which makes it easier to develop streaming applications. This component provides real-time data processing capabilities in Spark. It allows you to process real-time data streams, such as log files, social media feeds, and sensor data, using the same Spark programming model as batch processing.

    Spark Streaming works by dividing the real-time data stream into small batches and then processing each batch using Spark's batch processing engine. This approach allows Spark Streaming to provide fault-tolerant and scalable processing of real-time data streams. Spark Streaming provides a wide range of input sources, such as Kafka, Flume, HDFS, and Twitter. You can also create custom input sources by implementing a custom receiver that ingests data from a specific source. To process streaming data, Spark Streaming provides a set of high-level operations that are similar to those used in batch processing. These operations include map, reduce, filter, and join. You can also use machine learning algorithms to process streaming data and build predictive models in real-time.

    One of the advantages of Spark Streaming is that it can be seamlessly integrated with other Spark components, such as Spark SQL, Spark MLlib, and Spark GraphX. This allows you to perform complex analysis of streaming data and combine it with historical data stored in Spark.

    Spark Streaming provides several tools and libraries that make it easier to develop, monitor, and manage streaming applications. Here are some of the most commonly used tools:

    - ○ **Spark Streaming UI:** The Spark Streaming UI provides a web-based interface that displays real-time information about the status and performance of your streaming application. It allows you to monitor the progress of your application, track the number of records processed, and identify any errors or performance issues.

o **Spark Streaming Checkpointing:** Checkpointing is a mechanism that allows Spark Streaming to recover from failures by storing the state of the application periodically on a reliable storage system, such as HDFS. This ensures that your application can resume from the point of failure without losing any data.

o **Spark Streaming Receiver Reliability:** Spark Streaming provides a reliable receiver called the *receiver-based approach* that guarantees the delivery of data, even in the event of a node failure or network partition.

o **Spark Streaming Integration with other Spark components:** Spark Streaming can be easily integrated with other Spark components, such as Spark SQL, Spark MLlib, and Spark GraphX, to perform more complex analysis on the streaming data.

o **Spark Streaming Tuning:** Spark Streaming provides several configuration options that can be tuned to optimize the performance of your application. These include setting the batch interval, adjusting the number of executors, and increasing the memory allocated to the Spark driver and executors.

o **Spark Streaming Testing Libraries:** Spark Streaming provides testing libraries such as StreamingSuiteBase that allow you to test your streaming applications in a simulated environment.

Spark Streaming provides a comprehensive set of tools and libraries that make it easier to develop, monitor, and manage streaming applications.

- *Spark MLlib*: Apache Spark is a data processing framework that offers a scalable and distributed environment for the development and deployment of machine learning models. Spark MLlib is a machine learning library that was created on top of Apache Spark. It provides a wide range of machine learning algorithms and tools for data pre-processing, feature engineering, model training, and evaluation. It includes algorithms for classification, regression, clustering, and collaborative filtering, as well as tools for feature extraction and transformation.

Here are some of the key features of Spark MLlib:

o **Distributed and scalable:** Spark MLlib is designed to scale to handle large datasets and complex machine learning models. It distributes the data and computations across a cluster of machines, making it possible to process data in parallel.

○  **Variety of algorithms:** Spark MLlib provides a wide range of machine learning algorithms, including classification, regression, clustering, collaborative filtering, and dimensionality reduction. It also provides tools for feature engineering, such as feature extraction, transformation, and selection.

○  **Easy-to-use API:** Spark MLlib provides a high-level API that is easy to use and allows you to build and train machine learning models quickly.

○  **Integration with other Spark components:** Spark MLlib can be easily integrated with other Spark components, such as Spark SQL and Spark Streaming, to perform more complex analysis on the data.

○  **Model persistence:** Spark MLlib allows you to persist your models to disk or distributed file systems, such as HDFS, making it easier to deploy them in production.

○  **GPU support:** Spark MLlib now supports running some of the machine learning algorithms on GPUs for even faster processing of large-scale data.

Spark MLlib is a powerful machine learning library that makes it easy to build and deploy scalable and distributed machine learning models on top of Apache Spark. It is widely used in various industries, such as finance, healthcare, and e-commerce.

•  *Spark GraphX*: Spark GraphX is a distributed graph processing system built on top of Apache Spark. It provides a distributed and scalable framework for working with large-scale graphs and networks, such as social networks, web graphs, and biological networks. This component provides graph processing capabilities in Spark. It allows us to model and analyze complex relationships between entities in large datasets.

Here are some of the key features of Spark GraphX:

○  **Distributed and scalable:** Spark GraphX is designed to scale to handle large graphs and networks. It distributes the graph and computations across a cluster of machines, making it possible to process data in parallel.

○  **Graph construction:** Spark GraphX provides tools for building graphs from different data sources, such as CSV files, RDDs, and Parquet files.

○  **Graph transformation:** Spark GraphX provides a wide range of graph transformation operations, such as mapVertices, mapEdges, subgraph, and joinVertices, that allow you to modify the graph structure and properties.

o **Graph algorithms:** Spark GraphX provides a wide range of graph algorithms, such as PageRank, Connected Components, Triangle Counting, and Label Propagation, that allow you to perform various analyses on the graph.

o **Graph persistence:** Spark GraphX allows you to persist your graphs to disk or distributed file systems, such as HDFS, making it easier to reuse them in different applications.

Spark GraphX provides a scalable and distributed environment for graph processing and analysis, making it a popular choice for various industries, such as social media, e-commerce, and finance.

- *SparkR*: This component provides an R-based interface for working with Spark. It allows us to use the R programming language to manipulate and analyze Spark data.

The Spark ecosystem provides a comprehensive set of tools and capabilities for distributed data processing and analytics, including batch processing, real-time processing, machine learning, and graph processing.

As shown in Figure 6.6, various library modules are constructed on top of the Spark framework. They consist of the following:

- Spark Streaming for real-time streaming applications,
- GraphX for applications involving distributed graph processing,
- SQL for database applications and MLlib for distributed machine learning applications.

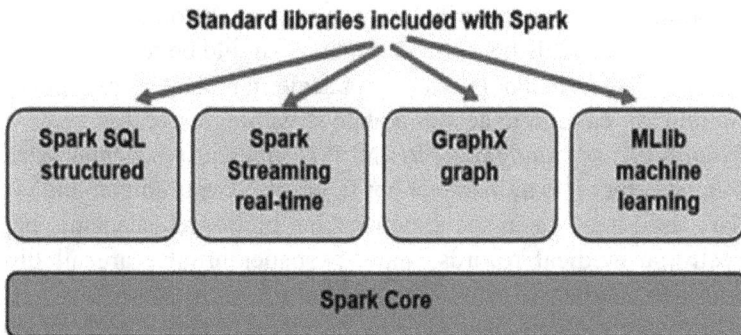

**Figure 6.6.** Spark Framework and its Components

This suggests that libraries atop the Spark framework can process applications for stream processing, graph processing, and machine learning.

## 6.3.1 Language Processing with Spark 2.0

On top of Apache Spark and Spark ML, Spark NLP is an open-source natural language processing library. It offers a simple API for ML Pipelines integration, and John Snow Labs offers commercial support for it. Rule-based algorithms and machine learning, some of which have Tensorflow operating in the background, are used by Spark NLP's annotators to power particular deep learning implementations.

The library includes a wide range of typical NLP tasks, such as named entity recognition, sentiment analysis, part-of-speech tagging, tokenization, stemming, and lemmatization. The online reference includes a description of the whole collection of annotators, pipelines, and concepts. Models can be trained using your data, and any of the models can be used since they are all open-source. Additionally, it offers pre-trained pipelines and models; however, these are only for demonstration purposes and not for usage in actual production.

Scala and Python APIs are also incorporated into the Scala-based Spark NLP library. It does not rely on any NLP or ML libraries to function properly. After conducting an academic literature analysis to ascertain the current state of the art for each variety of annotator, we gather as a team to make a decision regarding the algorithm(s) to implement. Three factors are used to assess implementations:

- *Accuracy*: A superb framework is useless if its algorithms or models are poor. Performance during runtime ought to be on par with or superior to any available benchmark. Nobody should be forced to sacrifice accuracy because annotators cannot scale in a cluster environment or run quickly enough to accommodate streaming use cases.
- *Trainability or Configurability*: NLP is inherently a domain-specific issue, whether it is trainable or configurable. The grammar and vocabulary used in posts on social networking platforms, academic papers, electronic medical records, and newspaper articles are all distinct from one another.
- In software systems that outgrow previous libraries, such as spaCy, NLTK, and CoreNLP, Spark NLP is designed for production use.

The library is the most extensively used NLP library by enterprise companies as of February 2019, with 16% of them using it.

The library offers straightforward, effective, and accurate NLP notations for machine learning pipelines that scale well in a distributed context. It is built natively on Apache Spark and TensorFlow. This library incorporates NLP capabilities while also utilizing the Spark ML pipeline.

O'Reilly has conducted an annual survey and found various patterns in the adoption of artificial intelligence by enterprise companies. The Spark NLP library was ranked as the seventh most popular AI framework and tool overall in the study results. It is also twice as popular than spaCy and by far the most widely used NLP library. It was also discovered to be the most well-liked AI library, surpassing TensorFlow, Keras, PyTorch, and Scikit-Learn.

## 6.3.2  Analysis of Streaming Data with Spark 2.0

Spark Streaming is an extension of the primary Spark API that allows for real-time data streams to be handled in a way that is scalable, has high throughput, and is fault-tolerant. Data can be ingested from a variety of sources, such as Kafka, Kinesis, or TCP connections, and then processed utilizing advanced algorithms that are specified via high-level functions, such as map, reduce, join, and window. Finally, it is possible to push processed data to databases, filesystems, and real-time dashboards, as shown in Figure 6.7. In reality, you can use data streams to apply Spark's machine learning and graph processing techniques.

**Figure 6.7.**   Spark Streaming

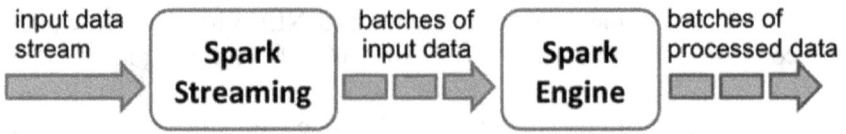

**Figure 6.8.**    Working of Spark Streaming

Internally, it functions as follows. Live input data streams are received by Spark Streaming, which divides the data into batches. The batches are then processed by the Spark engine to produce the final stream of results, as shown in Figure 6.8.

A continuous flow of data is referred to as a *discrete stream,* also known as DStream. Spark Streaming offers a high-level abstraction known as DStream, which is used to describe continuous data flows. Either by executing high-level operations on already existing DStreams or by absorbing raw data streams from sources such as Kafka and Kinesis, DStreams can be formed in one of two ways. An internal representation of a DStream is made up of a group of RDDs.

### 6.3.3  Streaming API

For users seeking accurate, current results, streaming APIs are used to receive data from the web in real time. For instance, social networking platforms often use them to distribute media content, such as audio and data. Generally, WebSocket — a subset of streaming APIs — is used by social networks. They obtain the data from servers, which then make them available for data consumption by software systems. Many businesses are utilizing this technology to get their data instantly. Facebook employs the technique to obtain real-time data updates from social graph subscriptions. Instagram is also involved because they use streaming APIs to get their real-time photo updates. This greatly improves the user experience for the consumer.

### 6.3.4  Kafka

Kafka [6,7] is actually a message broker. All of your data can pass through Kafka with excellent throughput before being sent to apps. A data pipeline is how Kafka operates. Generally, Kafka Stream enables millisecond-latency per-second stream processing.

Kafka Streams is a client library that allows for the processing and analysis of data that has been saved in Kafka [8,9]. Data can be processed in two different ways using Kafka streams:

- Kafka → Kafka: If designed properly, Kafka Streams delivers great scalability, high availability, high throughput, etc., when it conducts aggregations, filtering, etc., and publishes the data back to Kafka. Moreover, it does not perform "true streaming", or small batching.
- External Systems → Kafka (also known as "Kafka → Database" or "Kafka → Data science model"): Kafka is typically used as a message broker by any streaming library (including Spark, Flink, NiFi, etc.). After reading the messages from Kafka, it would divide them into smaller time frames so that it could continue processing them.

### 6.3.4.1 *Kafka Streaming*

Kafka streaming data can come from a variety of places, including event logs, website events, etc., as shown in Figure 6.9. Kafka streams, like those used by any other streaming application, would be used to access databases and models.

The proper separation of event time and processing time, the ability to create windowed views, and a straightforward (but efficient) approach to managing the application state are the pillars on which Kafka Streams is built. It is based on numerous ideas that are already present in Kafka, such as scaling through partitioning. It also comes in the form of a lightweight library that may be used in an application for this reason.

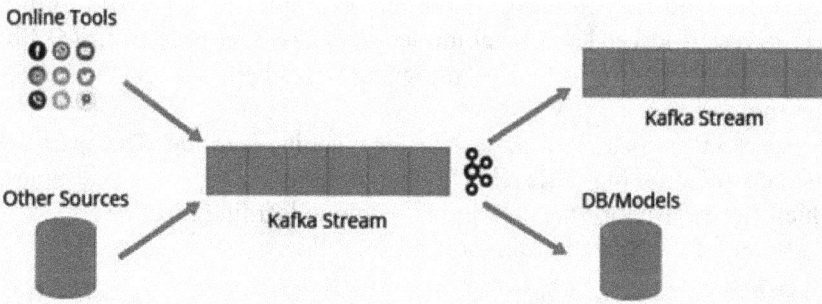

**Figure 6.9.** Kafka Streaming

**Figure 6.10.**   Apache Spark Streaming

### 6.3.5 Apache Spark Streaming

The data are divided into micro-batches by Spark Streaming, which receives live input data streams. The data are then processed by the Spark engine to produce the final stream of results in micro-batches. The operation of Spark Streaming is described in the data flow diagram shown in Figure 6.10.

Spark Streaming is built on top of the Spark core engine and provides a high-level API for working with data streams. The API provides functions for data input, processing, and output, and supports a wide variety of input sources. One of the key benefits of Spark Streaming is its ability to scale horizontally by adding more nodes to the Spark cluster. This enables it to handle large volumes of data and perform complex operations on it in real time.

## 6.4 Spark Machine Learning Library

Spark Machine learning is accomplished with Apache Spark by utilizing MLlib. The MLlib library contains many widely used tools and algorithms. This library is scalable, and it discusses both high-quality algorithms and high performance. Examples of machine learning algorithms include regression, classification, clustering, pattern mining, and collaborative filtering. Another example is pattern mining. Primitives of lower-level machine learning are also present in MLlib. One example of this would be the general gradient descent optimization method.

spark.ml is Spark's primary API for machine learning. The spark.ml package offers a higher-level API that is built on top of DataFrames, which can be used for the creation of machine learning pipelines.

Some of the Spark MLlib tools are as follows:

- *ML Algorithms*: ML algorithms serve as the library's primary building blocks. They include well-known methods of learning. such as clustering, regression, classification, and collaborative filtering. Standardizing interfaces is something that MLlib does in order to make it easier to include several algorithms in a single pipeline or process. One of the core ideas is the Pipelines API, and the Scikit-Learn project was the source of inspiration for the pipeline notion.

- *Featurization*: The process of featurization includes the steps of selection, dimensionality reduction, feature extraction, and feature transformation. The process of extracting features from unprocessed data is referred to as *feature extraction*. The process of resizing, updating, or otherwise altering features is what is meant by the term *feature transformation*. From among an extensive list of characteristics, only a select few of those that are most important are picked for inclusion in the final product.
- *Pipelines*: A pipeline connects a number of transformers and estimators together so that a machine learning workflow can be specified. In addition to this, it provides resources for constructing, evaluating, and fine-tuning machine learning pipelines. In the field of machine learning, it is common practice to execute a sequence of algorithms in order to process the data and gain knowledge from it. In MLlib, a process like this is represented by a pipeline, which consists of a sequence of pipeline stages (transformers and estimators) that need to be carried out in a specific order.
- *Persistence*: Persistence allows for the saving and loading of algorithmic models, pipelines, and other pipeline components. Because the model is permanent, it may be loaded or reused anytime it is required, saving both time and effort in the process.
- *Utilities*: Data processing, statistics, and linear algebra are the primary focuses of utilities' development. An illustration of this would be the MLlib tools for linear algebra, which are denoted by the filename "mllib.linalg".

## 6.5 Chapter Summary

In this chapter, we have seen how Spark uses robust distributed datasets to get around MapReduce's drawbacks. This enables simple and effective programming, allowing Spark to tune up to 100 times quicker than MapReduce operations. Spark eliminates the need for input/output operations by doing the majority of its operations over all iterations in memory instead of using a distributed file system. We also saw the different libraries that may be connected with Spark core, including GraphX for graph processing and MLlib for machine learning. Additionally, it is determined that using coarse-grained operations with Spark for failure recovery is not an overhead. A distributed messaging system with high performance in real time is called Kafka. Because it uses Zookeeper to achieve system fault tolerance, it is very dependable. Furthermore, since it is capable of horizontal scaling, Kafka offers greater scalability. In this chapter, we learned about a variety of applications in today's world, where a business has to comprehend a user's activity pattern, such as the information about mouse clicks that is collected from multiple data sources and aggregated. Then, utilizing the Kafka messaging system, Kafka will either perform a real-time analysis or store it in the backup. Distributed systems can use Kafka as an external commit log.

# References

1. Zaharia, M., Chowdhury, M., Franklin, M. J., Shenker, S., & Stoica, I. (2010). Spark: Cluster computing with working sets. *HotCloud*, 10(10), 95.
2. Zaharia, M., Chowdhury, M., Das, T., Dave, A., Ma, J., McCauley, M., Franklin, M. J., Shenker, S., & Stoica, I. (2012). Resilient distributed datasets: A fault-tolerant abstraction for in-memory cluster computing. In *Proceedings of the 9th USENIX Conference on Networked Systems Design and Implementation* (p. 2), 25–27 April 2012, San Jose, California.
3. Gonzalez, J. E., Xin, R. S., Dave, A., Crankshaw, D., Franklin, M. J., & Stoica, I. (2014). Graphx: Graph processing in a distributed data ow framework. In *11th USENIX Symposium on Operating Systems De-sign and Implementation (OSDI 14)* (pp. 599–613).
4. Armbrust, M., Xin, R. S., Lian, C., Huai, Y., Liu, D., Bradley, J. K., Meng, X., Kaftan, T., Franklin, M. J., & Ghodsi, A. (2015). Spark sql: Relational data processing in spark. In *Proceedings of the 2015 ACM SIGMOD International Conference on Management of Data* (pp. 1383–1394), 31 May–4 June 2015, Melbourne, Victoria, Australia. ACM.
5. Tripathi S., Gupta, B., Almomani, A., Mishra, A., & Veluru, S. (2013). Hadoop based defense solution to handle distributed denial of service (ddos) attacks, *Journal of Information Security*, vol 4, no. 3, DOI: 10.4236/jis.2013.43018.
6. Kreps, J., Narkhede, N., & Rao, J. (2011, June). Kafka: A distributed messaging system for log processing. In *Proceedings of the NetDB*, Vol. 11, No. 2011, pp. 1–7.

7. Dobbelaere, P. & Esmaili, K. S. (2017). Kafka versus Rab-bitMQ: A comparative study of two industry reference publish/subscribe implementations: Industry paper. In *Proceedings of the 11th ACM International Conference on Distributed and Event-based Systems* (pp. 227–238). ACM. Barcelona Spain.
8. Estrada, R. & Ruiz, I. (2016). *The Broker: Apache Kafka. Big Data SMACK* (pp. 165–203), Springer Nature, Switzerland AG.
9. Narkhede, N., Shapira, G., & Palino, T. (2017). *Kafka: The Definitive Guide: Real-time Data and Stream Processing at Scale.* O'Reilly Media, Inc. Sebastopol, California.

# Chapter 7

# Managing Big Data in Cloud Storage

Big data are large, voluminous datasets that require specialized tools and techniques to manage, process, and analyze. Cloud storage, on the other hand, refers to the practice of storing data on remote servers that can be accessed over the Internet.

Cloud storage is an ideal solution for storing and managing big data because it provides virtually unlimited storage capacity and can be accessed from anywhere with an internet connection [1]. Additionally, cloud storage providers typically offer a range of tools and services that are designed to support big data workloads, such as distributed computing frameworks, data processing engines, and analytics tools [2].

One of the key benefits of using cloud storage for big data is scalability. With cloud storage, organizations can quickly and easily scale their storage capacity up or down as their data needs change [3]. This can be especially useful for businesses that experience seasonal spikes in data usage or need to rapidly scale up their storage capacity in response to unexpected events. Another benefit of using cloud storage for big data is cost efficiency. Cloud storage facilitates customers a pay as per your usage pricing model, which means that businesses only pay for the storage capacity and services they actually use. This can be much more cost-effective than maintaining on-premises storage infrastructure, which can be expensive to purchase, maintain, and upgrade.

Managing big data in cloud storage involves several steps and considerations to ensure efficient storage and retrieval of large amounts of data [4]. The following points are key to managing big data in cloud storage:

- *Plan the data storage architecture*: Before storing big data in the cloud, one must carefully plan the storage architecture. There are various storage options, such as object storage, block storage, and file storage. Choose the storage option that suits your data type, usage pattern, and budget.
- *Choose a suitable cloud provider*: One should choose a cloud provider that offers storage and processing capabilities as per the requirements of big data. A few active cloud providers are Amazon Web Services (AWS), Microsoft Azure, and Google Cloud Platform.
- *Use data compression*: Compressing the data can save space and reduce costs in the cloud. Use compression tools, such as gzip, bzip2, or LZ4, to compress the data before uploading it to the cloud.
- *Use data partitioning*: Dividing data into smaller chunks and managing them definitely helps in performance improvement as well as reducing retrieval time. One can use partitioning techniques such as sharding, range partitioning, or hash partitioning.
- *Use cloud-native tools*: Many cloud providers offer cloud-native tools for managing big data, such as Amazon S3, Azure Blob Storage, and Google Cloud Storage. These tools provide features such as auto-scaling, data analytics, and real-time data processing.
- *Monitor and optimize performance*: Monitor big data storage and retrieval performance using cloud monitoring tools. Optimize performance by adjusting storage and retrieval parameters, such as network bandwidth, IOPS, and cache size.
- *Secure the data*: Ensure that big data is secure in the cloud by implementing security measures, such as encryption, access controls, and firewalls.

By following these tips, one can efficiently manage big data in cloud storage and take advantage of the scalability and flexibility of cloud computing.

## 7.1 Large-Scale Data Storage

Big data analytics facilitates better understanding and high utility analysis. However, it's quite difficult to handle its storage as it is more costly, needs scalability, and requires more data protection [5]. More data provides more information but at the expense of equally high scalable storage, which must be simple, reliable, and compatible with the other tools too.

Big data storage follows a storage-computation model where we collect and manage huge data volumes and perform real-time data analyses [6]. This informative data analysis can be used to mine associations and patterns that can be utilized in business intelligence from metadata. Typically, in real-world scenarios, this big data storage is composed of low-cost hard disk drives. However, nowadays, we are migrating toward flash storage, as it is in a phase of decreasing cost. When a flash storage is used, systems can be built purely on flash media or as hybrids of flash and disk storage.

Big data is formally unstructured data; therefore, object- and file-based storage are considered to build big data storage. In reality, these storages are not restricted to any particular capacities, and typically, volumes scale to a terabyte or petabyte size.

## 7.1.1 Challenges of Storing Large Data in Distributed Systems

Big data storage configuration and implementation encounter certain challenges as follows [7]:

- *Size and storage cost*: Storing large amounts of data in distributed systems can be expensive, as it requires a significant amount of storage capacity and computing resources.
- *Consistency*: In a distributed system, it can be difficult to ensure that all nodes have the same view of the data. Different nodes may have different versions of the data, and it can be challenging to synchronize these versions to ensure consistency.
- *Security*: Distributed systems must be able to ensure the security and privacy of the data being stored. This includes protecting against unauthorized access, ensuring data confidentiality, and providing data integrity.
- *High availability*: Distributed systems must ensure that data are available to users at all times, even in the event of a failure. This requires the system to have redundancy, fault tolerance, and failover mechanisms in place to ensure that data are always available, even when one or more nodes or servers fail.

- *Recovery*: Storing large amounts of data in a distributed system can be a significant challenge when it comes to recovery. Distributed systems are composed of multiple nodes that work together to store and process data, and any one of these nodes can fail at any time.

All these challenges arise in different scenarios, such as public clouds or in on-premises storage.

There are several platforms available for storing big data. Here are some popular ones:

- *Hadoop*: Apache Hadoop is an open-source framework for the distributed storage and processing of large datasets. It provides us with a distributed file system (HDFS) that can easily manage and store such a large data volume across multiple nodes in a cluster and a processing engine (MapReduce) that can parallelize computations on the data.
- *Amazon S3*: Amazon provides a S3 service, namely Simple Storage Service (S3), which is a cloud-based object storage service that provides highly scalable, durable, and secure storage for big data. It allows users to store and retrieve any amount of data from anywhere on the web.
- *Google cloud storage*: Google cloud storage is a scalable and widely available cloud storage service for big data. It provides a simple and flexible storage option for the purpose of storing and retrieving large amounts of data.
- *Apache cassandra*: It is a distributed NoSQL database that can store and manage large amounts of unstructured data. It is highly scalable and fault-tolerant, and can handle large datasets across multiple nodes in a cluster.
- *Microsoft azure blob storage*: A cloud-based object storage service that provides a solution that is highly scalable and cost-effective for the storage of big data. It facilitates users for storage and managing unstructured data in the cloud.

These platforms provide different features, pricing models, and performance characteristics, so choosing the right platform depends on the specific needs of the user and their big data requirements.

## 7.2  Hadoop Distributed File System (HDFS)

HDFS is a distributed file system that manages large volumes of datasets that can be stored or run on commodity machines [8]. An Apache Hadoop cluster can be scaled to hundreds (or even thousands) of machines, or, as we call them, nodes. HDFS is one of the major and most important components of Apache Hadoop. MapReduce and YARN are other important sub-components. HDFS is not Apache HBase, so it must not be confused with it; it is a column-oriented non-relational database management system that is modeled on top of HDFS. Using this, it actually better supports real-time data needs with its in-memory processing engine.

### 7.2.1  HDFS Permission Checks

Each HDFS operation clearly requires each user to have some specific permissions (such as READ, WRITE, and EXECUTE) via providing or granting ownership of, group membership, or other permissions [9]. The various activities carry out permission checks in order to verify and check other multiple components along the way, rather than focusing solely on the component that will ultimately be used. In addition, the owner of a path must be validated before certain activities can proceed.

Access to the traversal space is necessary for each and every one of the processes. It is exempt from this requirement for the final component of the path because it requires EXECUTE permission for all of the other components that exist on the path. For instance, the caller needs to have EXECUTE permission on /, /foo, and /foo/bar in order to carry out any operation that accesses the /foo/bar/baz directory.

Table 7.1 presents all the operations that are required by the HDFS for every component appearing in the whole path:

- *Ownership*: It determines whether or not a given caller is an owner in relation to the path by checking for this. In most cases, it is a reference to the actions that possess the authority to alter the ownership or permission metadata and demand that the caller be the owner.
- *Parent*: The directory that the requested path's parent directory is relative to. As an illustration, the parent of the path /foo/bar/baz is the directory /foo/bar.

- *Ancestor*: Ancestor refers to the component in the requested path that has been around the longest and was used most recently. As an illustration, the parent path for the route /foo/bar/baz is /foo/bar if the path /foo/bar already exists. If the route /foo exists but the path /foo/bar does not, then /foo is the ancestral path.
- *Final*: This is the very last part of the path that has been requested, and it is the component that has the most significance. For instance, the last component of the path /foo/bar/baz is the component /foo/bar/baz.
- *Sub-tree*: For each path that is already a directory, the sub-directories that are created for it and all of its children are listed in a recursive way. For instance, the sub-tree for the path /foo/bar/baz, which consists of two sub-directories named buz and boo, comprises the directories /foo/bar/baz, /foo/bar/baz/buz, and /foo/bar/baz/boo.
- When the call utilizes the overwrite option on the final path and there is already a file located at the path, WRITE access is necessary in order to avoid overwriting the file. Any operation that must determine whether or not it has permission to WRITE must also determine whether or not it has ownership, just in case the sticky bit is enabled.
- Calling set owner to change the user that owns a file requires HDFS super-user access. It is not necessary to have access as an HDFS super-user in order to modify the group, but the person making the call must be the owner of the file and a member of the group that was supplied.

Other permission options are presented in Table 7.1.

## 7.2.2  HDFS Shell Commands

A number of commands are shell-like and directly interact with the HDFS and also other file systems that support local file systems, HFTP File System, Special S3 File System, and others [11]. The FS shell is invoked by:

bin/hadoop fs <args>

Path URIs are accepted as parameters by every single FS shell command. The structure of a URI looks like this: scheme:/authority/path. The file

**Table 7.1.**   Permission Options in HDFS [10]

| Operation | Ownership | Parent | Ancestor | Final | Sub-tree |
|---|---|---|---|---|---|
| append | NO | N/A | N/A | WRITE | N/A |
| concat | NO [2] | WRITE (sources) | N/A | READ (sources), WRITE (destination) | N/A |
| create | NO | N/A | WRITE | WRITE [1] | N/A |
| createSnapshot | YES | N/A | N/A | N/A | N/A |
| delete | NO [2] | WRITE | N/A | N/A | READ, WRITE, EXECUTE |
| deleteSnapshot | YES | N/A | N/A | N/A | N/A |
| getAclStatus | NO | N/A | N/A | N/A | N/A |
| getBlockLocations | NO | N/A | N/A | READ | N/A |
| getContentSummary | NO | N/A | N/A | N/A | READ, EXECUTE |
| getFileInfo | NO | N/A | N/A | N/A | N/A |
| getFileLinkInfo | NO | N/A | N/A | N/A | N/A |
| getLinkTarget | NO | N/A | N/A | N/A | N/A |
| getListing | NO | N/A | N/A | READ, EXECUTE | N/A |
| getSnapshotDiffReport | NO | N/A | N/A | READ | READ |
| getStoragePolicy | NO | N/A | N/A | READ | N/A |
| getXAttrs | NO | N/A | N/A | READ | N/A |
| listXAttrs | NO | EXECUTE | N/A | N/A | N/A |
| mkdirs | NO | N/A | WRITE | N/A | N/A |
| modifyAclEntries | YES | N/A | N/A | N/A | N/A |
| removeAcl | YES | N/A | N/A | N/A | N/A |
| removeAclEntries | YES | N/A | N/A | N/A | N/A |
| removeDefaultAcl | YES | N/A | N/A | N/A | N/A |
| removeXAttr | NO [2] | N/A | N/A | WRITE | N/A |
| rename | NO [2] | WRITE (source) | WRITE (destination) | N/A | N/A |
| renameSnapshot | YES | N/A | N/A | N/A | N/A |
| setAcl | YES | N/A | N/A | N/A | N/A |
| setOwner | YES [3] | N/A | N/A | N/A | N/A |
| setPermission | YES | N/A | N/A | N/A | N/A |
| setReplication | NO | N/A | N/A | WRITE | N/A |
| setStoragePolicy | NO | N/A | N/A | WRITE | N/A |
| setTimes | NO | N/A | N/A | WRITE | N/A |
| setXAttr | NO [2] | N/A | N/A | WRITE | N/A |
| truncate | NO | N/A | N/A | WRITE | N/A |

system scheme for the local file system is file, while the HDFS file system scheme is hdfs. Both the scheme and the authority can be ignored. In the absence of a specific scheme, the one that is selected by default in the configuration will be applied. Either hdfs:/namenodehost/parent/child or just /parent/child can be used to refer to an HDFS file or directory, such as /parent/child.

The majority of the commands available in the FS shell have the same behavior as their corresponding Unix commands. The differences are broken down for each of the commands individually. The output is written to stdout, while the error information is written to stderr. HDFS shell provides a command-line interface to interact with files and directories on HDFS. The following are some of the commonly used HDFS shell commands:

**appendToFile**
The "appendToFile" command is not a built-in command in the HDFS shell. To append data to an existing file on HDFS, you can use the "hadoop fs -appendToFile" command. The syntax of the command is as follows:

hadoop fs -appendToFile [localSrc] [dst]

Here, [localSrc] is the path to the file on the local file system that you want to append to the [dst] file on HDFS. If the [dst] file does not exist, it will be created.

**cat**
The "cat" command can be used to display the contents of a file on HDFS. The "cat" command is similar to the "cat" command in Unix/Linux and can be used to print the contents of a file to the console.

The syntax of the "cat" command is as follows:

hadoop fs -cat [path/to/file]

Here, [path/to/file] is the path to the file on HDFS that you want to display. For example, to display the contents of a file named "data.txt" located in the "/user/hadoop" directory on HDFS, the following command can be used:

hadoop fs-cat /user/hadoop/data.txt

This command will print the contents of the "data.txt" file to the console.

If the file is compressed in the gzip format, you can use the "zcat" command instead of "cat" to display the contents of the file. The "zcat" command automatically decompresses the gzip file and displays the contents. For example, to display the contents of a gzip compressed file named "data.txt.gz" located in the "/user/Hadoop" directory on HDFS, one can use the following command:

hadoop fs-zcat /user/hadoop/data.txt.gz

This command will decompress the "data.txt.gz" file and print the contents to the console.

### chgrp
The hdfs dfs -chgrp command is used to change the group ownership of a file or directory in HDFS.

The syntax of the command is

hdfs dfs -chgrp <group> <path>

where <group> is the new group that you want to assign to the file or directory and <path> is the path of the file or directory that you want to change the group ownership for.

### chmod
The hdfs dfs -chmod command is used to change the permissions of a file or directory in HDFS.

The syntax of the command is

hdfs dfs -chmod [-R] <mode> <path>

where <mode> is the new permissions that you want to assign to the file or directory and <path> is the path of the file or directory that you want to change the permissions for. The -R option is used to apply the permissions recursively to all files and directories under the given path.

The <mode> parameter is a three-digit octal value that represents the permissions for the file or directory. Each digit corresponds to a different set of permissions: The first digit represents the permissions for the owner of the file or directory. The second digit represents the permissions for the

**Table 7.2.**    File Permissions Using CHMOD Command

| Bit combination | Permission |
|---|---|
| 0 | No permission |
| 1 | Execute |
| 2 | Write |
| 3 | Write and Execute |
| 4 | Read |
| 5 | Read and Execute |
| 6 | Read and Write |
| 7 | Read, Write, and Execute |

group that owns the file or directory. The third digit represents the permissions for all other users. Each digit is a combination of three bits, where the first bit represents the read permission, the second bit represents the write permission, and the third bit represents the execute permission.

Table 7.2 shows the values for each combination of bits.

For example, if you want to set the permissions of the file /user/hadoop/sample.txt to rw-r--r-, you can use the following command:

hdfs dfs -chmod 644 /user/hadoop/sample.txt

Note that we have the appropriate permissions to change the permissions of a file or directory in HDFS. Only the file owner or a user with super-user privileges can change the permissions of a file or directory in HDFS.

**chown**
The hdfs dfs -chown command is used to change the ownership of a file or directory in HDFS.

The syntax of the command is

hdfs dfs -chown [-R] <owner> [:<group>] <path>

where <owner> is the new owner that you want to assign to the file or directory, <group> is the new group that you want to assign to the file or directory (optional), and <path> is the path of the file or directory that you want to change the ownership for. The -R option is used to apply the ownership changes recursively to all files and directories under the given path.

**copyFromLocal**

The hdfs dfs -copyFromLocal command is used to copy files from the local file system to HDFS.

The syntax of the command is

hdfs dfs -copyFromLocal <localsrc> <dst>

where <localsrc> is the path of the file or directory in the local file system that you want to copy to HDFS and <dst> is the destination path in HDFS where you want to copy the file or directory to.

**copyToLocal**

The hdfs dfs -copyToLocal command is used to copy files from HDFS to the local file system.

The syntax of the command is

hdfs dfs -copyToLocal <src> <localdst>

where <src> is the path of the file or directory in HDFS that you want to copy to the local file system and <localdst> is the destination path in the local file system where you want to copy the file or directory to.

**count**

The hdfs dfs -count command is used to count the number of directories, files, and bytes used in HDFS.

The syntax of the command is

hdfs dfs -count [-q] [-h] <path>...

where <path> is the path of the file or directory for which you want to count the number of directories, files, and bytes used. You can specify multiple paths separated by spaces.

The -q option is used to display the quota information for the directories and files. The -h option is used to display the sizes in a human-readable format.

**cp**

The hdfs dfs -cp command is used to copy files or directories within HDFS or between HDFS and the local file system.

The syntax of the command is

hdfs dfs -cp [-p] <src> <dst>

where <src> is the path of the source file or directory that you want to copy and <dst> is the destination path where you want to copy the file or directory to. The -p option preserves the attributes and permissions of the source files and directories while copying.

One can use the hdfs dfs -cp command to copy files or directories within HDFS, between HDFS and the local file system, or between two different HDFS clusters.

For example, to copy a file named example.txt from the /user/hadoop directory in HDFS to the /user/hadoop/data directory in HDFS, you would use the following command:

hdfs dfs -cp /user/hadoop/example.txt /user/hadoop/data

### du
The hdfs dfs -du command is used to estimate the space used by the files and directories in HDFS.

The syntax of the command is

hdfs dfs -du [-s] [-h] <path>...

where <path> is the path of the file or directory for which you want to estimate the space used. One can specify multiple paths separated by spaces.

The -s option is used to display the total space used by the files and directories, instead of showing the space used by each file and directory separately. The -h option is used to display the sizes in a human-readable format.

### expunge
The hdfs dfs -expunge command is used to permanently delete files and directories in HDFS that were previously deleted but are still retained in the trash. When files or directories are deleted in HDFS, they are moved to a trash folder and are not immediately deleted from the file system. The hdfs dfs -expunge command can be used to empty the trash and permanently delete the files and directories that were deleted previously.

The syntax of the command is

hdfs dfs -expunge

The above command will permanently delete all the files and directories that were previously deleted and are still retained in the trash.

**get**
The hdfs dfs -get command is used to copy files from HDFS to the local file system.
The syntax of the command is

hdfs dfs -get <src> <dst>

where <src> is the path of the file in HDFS that you want to copy and <dst> is the path of the destination file in the local file system where you want to copy the file to.

**getmerge**
The hdfs dfs -getmerge command is used to concatenate files in HDFS and copy the result to the local file system.
The syntax of the command is

hdfs dfs -getmerge <src> <dst> [addnl]

where <src> is the path of the directory in HDFS containing the files to concatenate, <dst> is the path of the destination file in the local file system where you want to copy the concatenated result to, and the optional [addnl] parameter is used to indicate whether or not to add a newline character at the end of each file.

**getfacl**
The hdfs dfs -getfacl command is used to retrieve the access control list (ACL) for a file or directory in HDFS.
The syntax of the command is

hdfs dfs -getfacl <path>

where <path> is the path of the file or directory for which you want to retrieve the ACL. For example, to retrieve the ACL for a file named example.txt in the /user/hadoop directory in HDFS, you would use the following command:

hdfs dfs -getfacl /user/hadoop/example.txt

This will output the ACL for the file in the following format:

# file: /user/hadoop/example.txt
# owner: hadoop
# group: hadoop
user::rw-
user:user1:r--
group::r--
mask::r--
other::---

This output shows the owner, group, and permissions for the file. The first line starting with # file: indicates the path of the file. The second and third lines, starting with # owner: and # group: respectively, indicate the owner and group of the file. The subsequent lines starting with user:, group:, mask:, and other: indicate the permissions for each of these entities.

**setfacl**

The hdfs dfs -setfacl command is used to set the ACL for a file or directory in HDFS.

The syntax of the command is

hdfs dfs -setfacl -option <acl_specification> <path>

where <option> can be -m to modify the existing ACL, -x to remove the existing ACL, or -b to remove the existing ACL and set a default ACL. <acl_specification> is the new ACL to set, and <path> is the path of the file or directory for which you want to set the ACL.

The <acl_specification> can be specified in two formats: the standard POSIX format or the access control expression (ACE) format. The POSIX format is similar to the output of the hdfs dfs -getfacl command and specifies the permissions for the owner, group, and others, as well as any named users or groups. The ACE format is a more fine-grained format that allows you to specify permissions for specific users or groups.

**ls**
The hdfs dfs -ls command is used to list the contents of a directory in HDFS.

The syntax of the command is

hdfs dfs -ls <path>

where <path> is the path of the directory for which you want to list the contents.

**lsr**
The hdfs dfs -lsr command is used to recursively list the contents of a directory and all its sub-directories in HDFS.

The syntax of the command is:

hdfs dfs -lsr <path>

where <path> is the path of the directory for which you want to list the contents recursively.

**mkdir**
The hdfs dfs -mkdir command is used to create a new directory in HDFS.

The syntax of the command is:

hdfs dfs -mkdir <path>

where <path> is the full path of the directory you want to create in HDFS.

**moveFromLocal**
The hdfs dfs -moveFromLocal command is used to move a file or directory from the local file system to HDFS.

The syntax of the command is

hdfs dfs -moveFromLocal <local_path> <hdfs_path>

where <local_path> is the full path of the file or directory on the local file system that you want to move to HDFS and <hdfs_path> is the full path of the destination directory in HDFS.

**moveToLocal**

The hdfs dfs -moveToLocal command is used to move a file or directory from HDFS to the local file system.

The syntax of the command is

hdfs dfs -moveToLocal <hdfs_path> <local_path>

where <hdfs_path> is the full path of the file or directory in HDFS that you want to move to the local file system and <local_path> is the full path of the destination directory on the local file system.

**mv**

The hdfs dfs -mv command is used to move a file or directory from one location to another within HDFS.

The syntax of the command is

hdfs dfs -mv <source_path> <destination_path>

where <source_path> is the full path of the file or directory that you want to move and <destination_path> is the full path of the destination location.

**put**

The hdfs dfs -put command is used to copy a file or directory from the local file system to HDFS.

The syntax of the command is

hdfs dfs -put <local_path> <hdfs_path>

where <local_path> is the full path of the file or directory on the local file system that you want to copy to HDFS and <hdfs_path> is the full path of the destination directory in HDFS.

**rm**

The hdfs dfs -rm command is used to remove a file or directory from HDFS.

The syntax of the command is

hdfs dfs -rm <hdfs_path>

where <hdfs_path> is the full path of the file or directory that you want to remove from HDFS. The -rm command permanently deletes the file or directory from HDFS.

## rmr
The hdfs dfs -rmr command is used to recursively remove a directory and all its contents from HDFS.
   The syntax of the command is

hdfs dfs -rmr <hdfs_path>

where <hdfs_path> is the full path of the directory that you want to remove from HDFS. The -rmr command permanently deletes the directory and all its contents from HDFS.

## setrep
The hdfs dfs -setrep command is used to set the replication factor for a file or directory in HDFS.
   The syntax of the command is

hdfs dfs -setrep [-R] <replication_factor><hdfs_path>

where <replication_factor> is the desired replication factor (a positive integer) and <hdfs_path> is the full path of the file or directory that you want to set the replication factor for.

## tail
The hdfs dfs -tail command is used to display the last kilobyte of a file in HDFS.
   The syntax of the command is

hdfs dfs -tail <hdfs_path>

where <hdfs_path> is the full path of the file that you want to display the last kilobyte of. The -tail command applies only to text files and does not work with binary files.

**test**

The hdfs dfs -test command with different options is used to test various conditions on a file or directory in HDFS. Here are some examples:

To test if a file or directory exists in HDFS, you can use the following command:

hdfs dfs -test -e <hdfs_path>

To test if a file or directory does not exist in HDFS, you can use the following command:

hdfs dfs -test -z <hdfs_path>

To test if a file or directory is empty in HDFS, you can use the following command:

hdfs dfs -test -z <hdfs_path>

where <hdfs_path> is the full path of the file or directory that you want to test.

**text**

The hdfs dfs -text command is used to read a compressed or uncompressed text file stored in HDFS and display the content on the terminal.

The syntax of the command is

hdfs dfs -text <hdfs_path>

where <hdfs_path> is the full path of the file that you want to read and display on the terminal.

**touchz**

The hdfs dfs -touchz command is used to create a new, empty file in HDFS. The file will be created with a zero length and with the current time stamp as the modification time.

The syntax of the command is

hdfs dfs -touchz <hdfs_path>

where <hdfs_path> is the full path of the file that you want to create in HDFS.

## 7.2.3  Chaining and Scripting HDFS Commands

HDFS provides a set of command-line tools that allow you to interact with the file system [12]. In this section, we discuss how to chain and script HDFS commands.

*Chaining HDFS Commands*: We can chain multiple HDFS commands together to perform complex operations. For example, we can use the following command to create a directory and then copy a file to that directory:

hdfs dfs -mkdir /mydir && hdfs dfs -put myfile.txt /mydir

In this command, the && operator is used to chain two commands together. The first command creates a directory /mydir, and the second command copies the file myfile.txt to that directory.

*Scripting HDFS Commands*: We can also write scripts to automate HDFS operations. HDFS provides a shell interface called the Hadoop shell, which allows us to execute HDFS commands in batch mode. We can write scripts using this interface to perform operations on HDFS.

For example, we can create a script called copy_files.sh to copy all the files from one directory to another directory. The script could contain the following commands:

```
#!/bin/bash
input_dir=$1
output_dir=$2
hdfs dfs -ls $input_dir | awk '{print $8}' | while read filename ; do
    hdfs dfs -cp $filename $output_dir
done
```

In this script, the first two lines define the input and output directories. The hdfs dfs -ls command is used to list the files in the input directory, and the awk command is used to extract the filename. The while loop reads each filename and uses the hdfs dfs -cp command to copy the file to the output directory.

To run the script, the following command is executed:

./copy_files.sh /input /output

This would copy all the files from the /input directory to the /output directory.

## 7.2.4 Loading Data on HDFS

One can choose to upload their data in HDFS or an object store [13]. Data can be loaded into HDFS by using

* HDFS CLI
* WebHDFS REST API

For sensitive data, we should use a secure location that was previously created in HDFS. We can upload files to your home directory (/user/clsadmin). However, we should avoid uploading files to the /tmp directory.

### *Uploading data by using the HDFS CLI*
To use the HDFS CLI, we must use SSH to log in to the cluster using the credentials.

**Prerequisites:** We need the user credentials and the SSH endpoint.

We access the HDFS CLI by using the HDFS command. Refer to the following examples for using the HDFS CLI:

* To create a directory under the user home:

```
hdfs dfs –mkdir –p /user/clsadmin/test-dir
```

* To upload a file to an existing HDFS directory:

```
hdfs dfs –put test-file /user/clsadmin/test-dir
```

* To delete a file or directory from HDFS:

```
hdfs dfs –rm –f /user/clsadmin/test-dir
```

## *Uploading data by using the WebHDFS REST API*

For programmatic access to the HDFS, use the WebHDFS REST API.

**Prerequisites:** We need user credentials and the WebHDFS URL.
To upload data to HDFS by using the WebHDFS REST API:

* Open a command prompt.
* Change the directory to the location of the data files that you want to upload. Using the WebHDFS URL that is identified by checking the docs under Prerequisites, make a REST API call by using the cURL command to show directory contents, create directories, and upload files.

For example, to show the current contents of your cluster's top-level HDFS directory, which is named /user, run the following command:

```
curl -i -s --user clsadmin:your_password --max-time 45 \
https://XXXXX:8443/gateway/default/webhdfs/v1/user?op=LISTSTATUS
```

The value of XXXXX is the host name of your cluster retrieved from the service endpoints JSON output. If the call completes successfully, 200 OK is returned.

* To upload a file, run the following command:

```
curl -i -L -s --user clsadmin:your_password --max-time 45 -X PUT -T
file_name.txt \
https://XXXXX:8443/gateway/default/webhdfs/v1/user/clsadmin/path_
to_file/file_name?op=CREATE
```

If the directories do not exist, they are created. If the call completes successfully, 201 Created is returned.
Run one cURL command for each file that you want to upload.

* To create an empty directory, for example an output directory, run the following command:

```
curl -i -s --user clsadmin:your_password --max-time 45 -X PUT
https://XXXXX:8443/gateway/default/webhdfs/v1/user/clsadmin/
path_to_directory?op=MKDIRS
```

- To remove a file, run the following command:

```
curl -i -s --user clsadmin:your_password --max-time 45 -X DELETE
https://XXXXX:8443/gateway/default/webhdfs/v1/user/clsadmin/
path_to_file?op=DELETE
```

You can't remove a directory that isn't empty.

An alternative way to look at the directory structure, contents, owners, and size is to navigate to the following URL:

```
https://<changeme>:8443/gateway/default/hdfs/explorer.html
```

where <changeme> is the URL to the cluster.

For example, for data on a cluster in the US-South region, use

```
https://XXXXX.us-south.ae.appdomain.cloud:8443/gateway/default/
hdfs/explorer.html
```

To load data onto HDFS, we first need to prepare the data by ensuring that it is in the appropriate format and structure. HDFS supports various file formats, such as text, sequence files, and Avro. Second, we need to connect to the Hadoop cluster using a command-line interface or a graphical user interface, such as Ambari or Cloudera Manager. Third, we need to create a directory on HDFS to store your data. One can use the Hadoop command-line interface to create a directory using the hdfs dfs -mkdir command. Fourth, copy data to HDFS using the hdfs dfs -put command. This command copies data from a local file system to the HDFS file system, and finally, verify that the data are loaded successfully by running the hdfs dfs -ls command to list the contents of the directory.

## 7.3  Hadoop User Experience (HUE)

Hue is an open-source interface for the user and Hadoop components. The user can easily access it directly from the browser in order to enhance the productivity of the Hadoop developers. It was initially developed and designed by Cloudera [14,15]. Through Hue, the user can interact with HDFS and MapReduce applications. The advantage lies in the fact that the users need not have the command-line interface in order to avail themselves of the benefits of the Hadoop ecosystem if they need to use HUE.

## 7.3.1 Features of HUE

Hue provides several important features and is not restricted to just a web interface for Hadoop developers. Hue provides a lot of features which make it a popular tool among Hadoop developers. These features are as follows:

- Hadoop API Access
- Presence of HDFS File Browser
- Browser and Job Designer
- User Admin Interface
- Editor for Hive Query
- Editor for Pig Query
- Hadoop Shell Access
- Workflows can access Oozie Interface
- SOLR searches can get a Separate Interface

The above-mentioned features encourage developers to make Hue a primary choice and are used in Hadoop cluster installation. All prime features of Hadoop features can be easily availed of through Hue, and clients or others who do not hold any familiarity with the command-line interface, i.e., cmd, can easily access Hue along with its functionality.

## 7.3.2 HUE Components

Hue consists of a number of components, as shown in Figure 7.1. These components deeply help users take advantage of the entire Hadoop eco-system, and Corley helps them implement it properly:

- *HDFS Browser*: Browser can be viewed as an interface for the purpose of managing and browsing the whole batch of files stored in the distributed file system, i.e., HDFS. When working with the Hadoop Ecosystem, one of the prime advantages is the access ability of the HDFS Browser, which actually helps users interact with the HDFS files in a closer way. The HDFS interface provides the privilege of

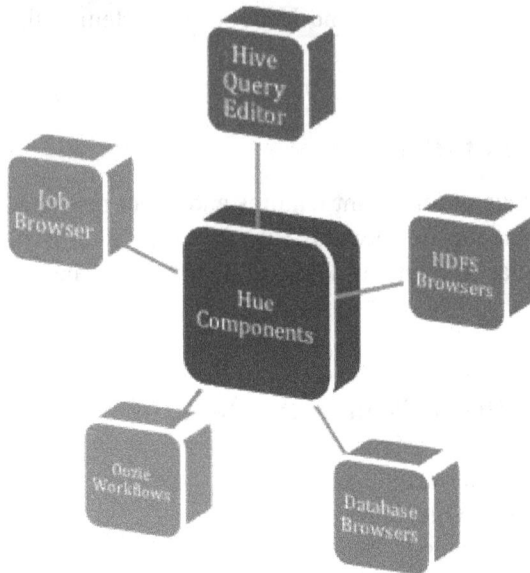

**Figure 7.1.**   Hue Components [14]

using the required operations that can be performed on the HDFS architecture. If a user is not comfortable with the command-line interface, then it provides an easy way to access it.

In the Hue interface, "File Browser" is present at the top-right position, as shown in Figure 7.2.

A file browser will be opened through the "File Browser" link. For the current or default path, it will list all the files along with their file properties. The user can even delete, download, or upload new files from here.

- *Job Browser*: This is a component that allows users to view and manage jobs running on the Hadoop cluster, including MapReduce jobs and other Hadoop applications. Hadoop ecosystems carry a number of jobs, and many times, the developer may require to see which job is currently in the running state on the Hadoop cluster and also which particular job has been successfully completed and which has errors. Through Job Browser, we can access all the job-related information right from inside the browser. For this, there is a button in Hue that can list the number of jobs and their status.

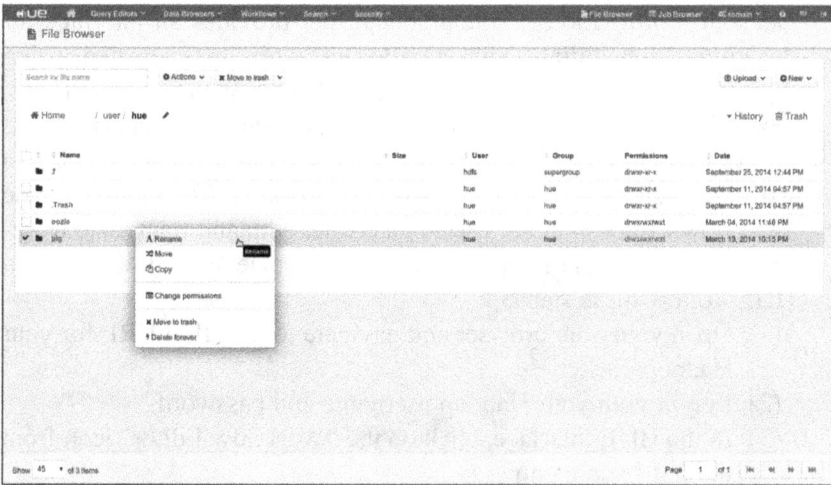

**Figure 7.2.**   Hue Interface

- *Hive Query Editor*: Hive Query Editor in HUE is a web-based graphical user interface that allows users to interact with Apache Hive, a data warehousing infrastructure built on top of Hadoop [16]. With the Hive Query Editor in HUE, users can write and execute Hive queries using a visual interface. To use the Hive Query Editor in HUE, follow these steps:
  (1) Open your web browser and navigate to the HUE URL for your Hadoop cluster.
  (2) Log in with your Hadoop username and password.
  (3) In the HUE interface, click on the "Query Editor" icon from the left-hand menu.
  (4) From the dropdown menu on the left, select "Hive" as the query editor type.
  (5) Write your Hive query in the editor on the right-hand side.
  (6) Click on the "Execute" button to run your query.
  (7) View the results of your query in the table below the editor.

  In addition to executing Hive queries, the Hive Query Editor in HUE also provides features such as syntax highlighting, autocomplete, and query history.

- *Oozie Workflow Editor*: A component that provides an interface for designing, scheduling, and monitoring workflows created using Apache Oozie, a system for managing Hadoop jobs and workflows. Oozie Workflow Editor is a web-based graphical user interface that allows users to create and manage workflows using Apache Oozie, a workflow scheduler system for Apache Hadoop [17]. With the Oozie Workflow Editor in HUE, users can create, edit, and run Oozie workflows using a visual interface. To use the Oozie Workflow Editor in HUE, follow these steps:
  (1) Open your web browser and navigate to the HUE URL for your Hadoop cluster.
  (2) Log in with your Hadoop username and password.
  (3) In the HUE interface, click on the "Workflow Editor" icon from the left-hand menu.
  (4) Click on the "Create" button to create a new workflow or select an existing workflow from the list.
  (5) Drag and drop actions from the left-hand side panel onto the canvas to build your workflow.
  (6) Configure the properties for each action by clicking on the action and filling out the form on the right-hand side panel.
  (7) Connect the actions in the workflow by dragging the output of one action to the input of another.
  (8) Click on the "Save" button to save your workflow.
  (9) Click on the "Validate" button to check your workflow for errors.
  (10) Click on the "Submit" button to run your workflow.
  In addition to creating and running workflows, the Oozie Workflow Editor in HUE also provides features such as syntax highlighting, autocomplete, and error checking. Users can also view the status and logs of their running workflows in the HUE interface.
- *Database Browsers*: One of the features of Hue is the database browser, which allows users to browse and manage databases and tables in their Hadoop cluster. There are several database browsers available in Hue, including:
  (1) Metastore Browser: This browser allows users to browse and manage metadata for tables and partitions in Hive, a data warehousing infrastructure built on top of Hadoop.

(2) HBase Browser: This browser allows users to browse and manage tables and column families in HBase, a distributed NoSQL database that runs on top of Hadoop.

(3) Impala Browser: This browser allows users to browse and manage tables and partitions in Impala, a massively parallel processing (MPP) SQL query engine for Hadoop.

(4) SQL Editor: This editor allows users to execute SQL queries on their data stored in Hadoop, with support for various database systems, such as MySQL, Oracle, and PostgreSQL.

In addition to browsing and managing databases and tables, the database browsers in Hue also provide features such as searching, filtering, sorting, and data visualization. Users can also perform actions such as creating tables, editing table schemas, and importing/exporting data from the browser interface.

## 7.4 Chapter Summary

In this chapter, we presented the different primary factors that can be used to manage big data in the cloud using HDFS and Hue. It involves setting up a Hadoop cluster in the cloud and storing data in HDFS, a distributed file system that can store petabytes of data across multiple nodes in the cluster. Hue provides a web-based interface that allows users to interact with various Hadoop tools and services. Organizations can use Hue to manage databases and tables, execute SQL queries, and create and manage workflows using Oozie. In addition, Hadoop clusters in the cloud can be scaled up or down as needed, allowing organizations to adjust to changing data storage and processing requirements. Overall, managing big data in the cloud using HDFS and Hue can provide organizations with the flexibility, scalability, and cost-effectiveness they need to handle large volumes of data.

## References

1. Yang, C., Huang, Q., Li, Z., Liu, K., & Hu, F. (2017). Big Data and cloud computing: Innovation opportunities and challenges. *International Journal of Digital Earth*, 10(1), 13–53.
2. Psannis, K. E., Stergiou, C., & Gupta, B. B. (2018). Advanced media-based smart big data on intelligent cloud systems. *IEEE Transactions on Sustainable Computing*, 4(1), 77–87.

3. Gupta, B. B., Agrawal, D. P., Yamaguchi, S., & Sheng, M. (2018). Advances in applying soft computing techniques for big data and cloud computing. *Soft Computing*, 22, 7679–7683.
4. Chang, V. & Wills, G. (2016). A model to compare cloud and non-cloud storage of Big Data. *Future Generation Computer Systems*, 57, 56–76.
5. Reiss, G. & Hütten, A. (2005). Applications beyond data storage. *Nature Materials*, 4(10), 725–726.
6. Padgavankar, M. H. & Gupta, S. R. (2014). Big data storage and challenges. *International Journal of Computer Science and Information Technologies*, 5(2), 2218–2223.
7. Galanis, L., Wang, Y., Jeffery, S. R., & DeWitt, D. J. (2003, January). Locating data sources in large distributed systems. In *Proceedings 2003 VLDB Conference* (pp. 874–885), 9–12 September 2003, Berlin, Germany. Morgan Kaufmann.
8. Borthakur, D. (2008). HDFS architecture guide. *Hadoop Apache Project*, 53(1–13), 2.
9. Warren, J. & Marz, N. (2015). *Big Data: Principles and Best Practices of Scalable Realtime Data Systems*. Simon and Schuster, Manhattan, New York, USA.
10. Jam, M. R., Khanli, L. M., Javan, M. S., & Akbari, M. K. (2014, October). A survey on security of Hadoop. In *2014 4th International Conference on Computer and knowledge Engineering (ICCKE)* (pp. 716–721), 29–30 October 2014, Mashhad, Iran. IEEE.
11. Borthakur, D. (2010). "HDFS Architecture", Document on Hadoop Wiki. http://hadoop.apache.org/common/docs/r0.
12. "Hadoop Shell Command Guide — Command Chaining", https://hadoop.apache.org/docs/current/hadoop-project-dist/hadoop-common/FileSystem Shell.html#Command_Chaining.
13. "Hadoop Tutorial: Using HDFS Commands with Examples", https://hortonworks.com/tutorial/hadoop-tutorial-getting-started-with-hdp/section/3/.
14. "Cloudera Documentation on HUE: 'HUE User Guide'" https://docs.cloudera.com/documentation/enterprise/latest/topics/hue.html.
15. "Apache Hadoop documentation on HUE: 'Apache Hadoop — HUE User Guide'", https://hadoop.apache.org/docs/current/hadoop-project-dist/hadoop-hdfs/WebHDFS.html#Hue_User_Guide.
16. Capriolo, E., Wampler, D., & Rutherglen, J. (2012). *Programming Hive: Data Warehouse and Query Language for Hadoop*. O'Reilly Media, Inc. Sebastopol, California.
17. Islam, M. K. & Srinivasan, A. (2015). *Apache Oozie: The Workflow Scheduler for Hadoop*. O'Reilly Media, Inc. Sebastopol, California.

# Chapter 8

# Big Data in Healthcare

Companies acquire large amounts of unstructured or semi-structured data for use in sophisticated analytics tools, such as predictive modeling and machine learning. These data are referred to as *big data*. Traditional software cannot handle the complexity of big data collections. Douglas Laney provided the definition that has gained the most traction and acceptance. Volume, velocity, and variety — collectively referred to as the "3 V's" — were the three dimensions in which Laney saw that (big) data was expanding [1]. The term *big data* itself alludes to the enormous size of such datasets. Big data is defined not just by its volume but also by its speed and variety. Data collected by a business or system may include transaction records, videos, audio recordings, written documents, and even log files, all of which fall under the category of variety. Velocity is the rate at which information is gathered and made available for analysis. Big data is now typically defined by these three V's. Almost all research fields, whether they are related to business or academia, are producing and studying big data for a variety of reasons. To understand the meaning of this massive volume of data, it would be necessary to employ revolutionary fusion and artificial intelligence (AI) techniques. Using machine learning (ML) techniques, such as neural networks and other forms of AI, for the tasks of analysis and decision-making would be a significant step forward. Yet, without the right hardware and software, big data might be hard to interpret. In order to effectively analyze this "endless sea" of data and generate actionable insights, we need to develop smarter web services and better methods for handling the data. With the correct data storage and analysis tools, big data can be utilized to increase the effectiveness of vital parts of society's infrastructure, such as healthcare, security, and communication.

## 8.1 Digitalization in Healthcare Sector

Digital health is the combination of medical expertise and IT programs or technologies to enhance patient care and oversight. As a result, it is now possible to use a smartphone to continuously check whether a patient has taken prescribed medications, monitor vital signs (such as pulse, blood pressure, and oxygen saturation), and even identify whether a patient at home has fallen down by tracking movement patterns and body temperature.

The following should benefit from the use of digital health technologies:

- the elderly allowed to stay longer in their comfortable social milieu instead of relocating to an old age home or care facility,
- improved patient compliance with treatment behavior,
- avoidance of unnecessary hospital admissions,
- a prevention-oriented way of life.

Electronic health records (EHRs) and electronic medical records (EMRs) both document the standard clinical and medical data collected from patients. Healthcare information elements, such as EHRs, personal health records (PHRs), and medical practice management software (MPM), have the potential to improve the quality and efficiency of health-care delivery while decreasing costs. The management and application of such medical records are increasingly dependent on technological development. Well-being monitoring tools and software platforms that can issue warnings and convey patient health information to relevant health-care practitioners have been rapidly developed and deployed thanks to a real-time biomedical and health monitoring system. These devices provide massive volumes of data that can be analyzed to provide real-time clinical or medical therapy [2]. The medical industry's big data has the potential to improve health outcomes and cut costs.

### 8.1.1 Use of Big Data in Medical Care

In the healthcare domain, the term *big data* refers to the massive volumes of data generated by the digitization of formerly infeasible tasks, such as the collection of patient records and the control of healthcare quality. The use of big data analytics in healthcare has many positive consequences, some of which may potentially save lives. At its core, *big data* refers to the massive amounts of information contributed by digitization, which is then gathered and processed using a specific technology. When used in healthcare, it will make use of accurate health data about a community (or an individual) and may help prevent epidemics, treat diseases, reduce expenditures, etc. Big data in healthcare comes from a variety of places, including EHRs, payer records, smart devices, and genetic databases [3,4]. Nonetheless, the number of resources available to doctors for gathering additional information about their patients is always growing. This information often comes in a wide variety of formats, making it difficult to use for most people. What matters now is not the quantity of data but rather its intelligent upkeep. The following healthcare-related big data sources provide for rapid and insightful information extraction when equipped with the right technology.

## 8.2 Big Data in Public Health

Due to the availability of different data measurement techniques, storage technologies, etc., a significant quantity of information is available and accessible for public health decision-making [5] and study. Many researchers have indeed emphasized the significance of big data within the healthcare sector [6], epidemiology [7], surveillance [8], etc.

### 8.2.1 Big Data Surveillance Using Machine Learning

Public health monitoring systems keep an eye on trends in disease incidence, health behavior, and environmental variables in order to distribute resources to sustain healthy populations [9]. All five types of big data have the ability to help authorities comprehend the public health situation, not just effluent data, which is one of the most well-known applications of big data for monitoring (such as Google Flu Trends). However, there are also analytical difficulties because of how big these new sources of data are.

In cases where prediction rather than testing is the analytical goal, the employment of machine learning models in the data science community has helped alleviate the "curse of dimensionality" associated with big datasets.

Machine learning, in its broadest sense, refers to methods that algorithmically train models by identifying data patterns. These methods fall into one of three categories: semi-supervised learning, unsupervised learning, or supervised learning. Identifying relationships between predictors and outcomes and maximizing prediction are the main characteristics of supervised learning. A supervised learning method might, for instance, be a regression model, while unsupervised learning makes use of the natural characteristics of the input data to discover trends without tagging any one column as the target result. Principal component analysis is an example of an unsupervised technique that can identify hidden covariance trends within data. Semi-supervised learning, a hybrid approach, is utilized when making predictions as the main objective, but most data points lack outcome information. Semi-supervised and unsupervised techniques are frequently employed as a prelude to supervised techniques meant for prediction or more thorough statistical analysis in a follow-up during the data mining phase. Some scholars and practitioners in public health have explored machine learning, even though data science has more extensively accepted it [10]. Unsupervised learning has, for instance, been used for monitoring systems, disease outbreak diagnosis and tracking [11], the identification of patient traits related to clinical outcomes [12], and spatiotemporal profiling [13]. To detect falls from smartphone data [14], construct early warning systems for hazardous medication responses, and identify outlier air pollutants [15], among other uses, semi-supervised variations of current machine learning models have been used. Among many other uses, supervised learning has been applied in numerous settings, including hospitalization prediction [16], the spread of TB [17], the severity of injuries sustained in car accidents [18], and the inclination of Reddit members toward suicide thoughts [19].

## 8.2.2 Big Data in Public Health Training

Big data usage in public health research and practice necessitates the development of new management and analytical skills, but it doesn't eliminate the need for other abilities that are typically associated with global health education, including domain expertise, governance, and statistical principles [20]. The training and efforts necessary to acquire and keep an up-to-date understanding of the latest developments in computational and analytical systems, however, are not straightforward. All big data practitioners may need to develop two key abilities. While interacting with data, it could be crucial to first learn how to think like a machine. Such computational thinking, whereby an analyzer can identify which situations represent higher technical hurdles, extends deeper than merely understanding programming techniques, operating software, or creating equipment and has been presented as a replacement for traditional early education subjects, such as reading, writing, and arithmetic [21]. However, when presented with large datasets that require a lot of time and resources, even global health professionals without a background in computing may find it useful to be capable of thinking like a machine. Second, statistical bias research and similar methods will probably play a bigger role in public health education, particularly in the fields of epidemiology and biostatistics. Investigations are predicted to employ secondary data more frequently as integrated, sophisticated, and large global health datasets become available. To maintain trust in substantiative conclusions, methods that can discover the possibility of inaccurate implications under various assumptions of partiality will be crucial [22]. In circumstances where an examiner was not involved in the process of data collection, it is significantly more difficult to disprove the existence of systematic biases. When making judgments regarding method selection and evaluation, you will also encounter conflicts similar to those that exist between the accuracy of specific data points and the probabilistic conceptions of correctness applied to the entire dataset. These two fundamental competencies only represent a portion of the total data science competencies required to deal with big data in the healthcare system, which also include comprehension of medical informatics, data engineering, computation cost, and active learning [23]. Nevertheless, because mastering these skills necessitates a significant financial investment, we claim that, for public well-being, teenagers ought to have access to workshops in more sophisticated data science methodologies, but it shouldn't be a requirement, similar to other elective but crucial skills, such as society health assessment [24]. Diverse teams will be required for this development of specialized abilities; public health professionals are already familiar with this paradigm, but it has not been fully implemented in training up to this point. Big data expertise should be leveraged to stream-line data-collection processes since both specialist and generalist big data skills are becoming prevalent in the public health sector [25]. A biostatisti-cian who is accustomed to actual data handling might be more likely to advocate for information trial protocols, whereas an informatics expert with knowledge of natural language might be able to help a practitioner write her memos in a way that will be most useful for both clinical and research purposes. The likelihood that epidemiologists who are familiar with stepped-wedge designs [26] will recommend them to decision-makers implementing public health programs may be higher. In general, acquiring new techniques for handling data efficiently will and ought to influence not only the data we want to gather but also the methods we employ.

### 8.2.3 Limitations and Open Issues for Big Data While Using Machine Learning in Public Health

Understanding a few of each technology's most important constraints is necessary for the effective usage of both big data and machine learning. We note first that large datasets are necessary for machine learning to escape the curse of dimensionality [27]. Overfitting can result from small, biased, or incomplete training sets, thereby restricting the range of issues that can be solved by present machine learning techniques. Second, machine learning frameworks are sometimes referred to as *black boxes*, whose opacity prevents non-experts from being able to analyze them or verify their fundamental premises. Third, some analysts believe that dynamically learning algorithms are much more precise than human-built models because they are more unbiased. Although theory-driven models are frequently able to provide more accurate predictions of outcomes than data-driven models, constructing data-driven models still requires making subjective decisions regarding the training and assessment datasets, pre-processing standards, learning algorithms, and initial parameters. As a result of all of these choices, biases and preconceptions may exist that are hidden from casual users. Last but not least, since big data studies com-monly necessitate trying to link supplementary data drawn from a variety of resources, inconsistencies between such resources can result in biases, including those with demographic patterns.

## 8.3 The Four V's of Big Data in Healthcare

Big data has swiftly become an indispensable resource for virtually every clinical and administrative action in the medical field, including managing population health, monitoring systems, revenue cycle administration, predictive modeling, and clinical decision-making. These are just some of the activities. Although it is challenging to boil down the complexity of big data analytics into manageable chunks, the dictionary has done a fantastic job of giving commentators some useful terms. Both data scientists and tech writers enjoy patterns, and few are more appealing to both occupations than the alliterative qualities of the many V's in big data. Until big data became a popular phrase, there were just the big three at first: volume, velocity, and variety, which were first suggested by Doug Laney, a Gartner analyst, earlier in 2001, as shown in Figure 8.1. IBM was crucial in introducing the fourth V, veracity, to the mixture as businesses began to gather an increasing variety of data types, some of which were missing or

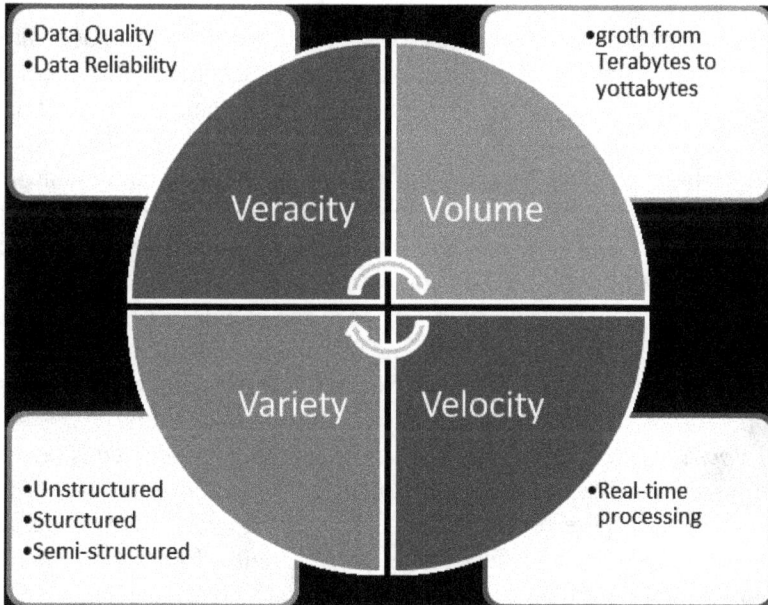

**Figure 8.1.** The Four V's of Big Data

poorly designed. Even more terms have been added to the litany as a result of subsequent linguistic advances. The following concepts have been suggested as possible additions to the list: value, visualization, viability, etc. Each word refers to a certain characteristic of big data that companies need to comprehend and manage to be successful with their selected efforts:

- *Volume*: It is hardly unexpected that big data is huge in bulk. It is estimated that humans generate 2.3 trillion gigabytes of data each and every day, and things are just going to get worse. Obviously, the vast mobile phone system is a component of this increase, as it is one of the contributing factors. To get an idea, six out of the seven billion individuals who currently inhabit the world have access to a cell phone. Pictures, videos, SMS, WhatsApp conversations, and connections with other apps are all factors that contribute to the massive expansion in the amount of data used. As the volume continues to rise at a rapid rate, there will be an increased demand for new database management solutions, as well as for additional IT personnel. It is projected that during the next few years, millions of new jobs in the field of information technology will be created in order to accommodate the onslaught of big data. Data from healthcare providers are almost always going to be more beneficial than the most recent Netflix binge. Information-dense clinical notes, claims data, laboratory findings, gene sequencing, data from medical equipment, and imaging tests become even more valuable when linked in novel ways to deliver brand-new insights. In order for enterprises to effectively manage the quantity of data at their disposal, they need to create storage strategies, either on-premises or in the cloud. In addition to this, they are responsible for ensuring that their infrastructure is able to support the upcoming "V" on the list without hindering essential operations, such as communications with their suppliers.
- *Velocity*: The rate at which data are produced and processed is referred to as their velocity, which is also sometimes referred to as speed. It wasn't until quite recently that sufficient time was spent processing the relevant data and uncovering the relevant information. Data are now readily available in real time. In addition to the speed of the internet, this is a consequence of the existence and presence of big data. Because we generate more data, we need more means to monitor them, and as a result, we monitor more data. This results in the

formation of a vicious circle. The quantity of data pertaining to healthcare that is transmitted over the wires of the world will increase as a result of the development of revolutionary data production and processing methods, such as the Internet of Things, medical devices, genetic testing, and machine learning.

- *Variety*: The numerous kinds of data are linked to the high speed and substantial amount of information. In the end, there are intelligent information technology solutions available today for every area, from the business world and the medical field to the construction industry and even the domestic sphere. Consider the computerized patient records utilized in the medical field. These records account for many billions of terabytes of data. And this doesn't even take into account the films that we watch on YouTube, the posts that we make on Facebook, or the articles that we write for our blogs. The quantity, as well as the variety, will continue to expand up until the point that internet access is available everywhere in the world. It is unfortunate that the unchecked expansion of information technology over a protracted period of time has resulted in many providers having data silos that are extremely difficult to penetrate, which is bad news for the healthcare industry. It is not practical to compare datasets that are kept in multiple locations or that utilize different formats. This prevents providers from gaining as much knowledge as they could otherwise about their customers or business processes. The range of incompatible data formats, semi-data structures, and inaccurate data semantics will be the biggest obstacles to efficient data management, according to Laney. Even though he was predicting the year 2004, his remarks still hold today. Application software interfaces (APIs) and emerging standards, such as FHIR, which make it simpler to overcome walled gardens and increase diversity, are helping healthcare IT developers begin to deconstruct the issue.
- *Veracity*: The veracity of big data is still an issue that's up for debate. Due to the rapid pace at which data become obsolete, the information that is shared on the internet and through social networks does not necessarily need to be accurate all of the time. The corporate world is full of directors and managers who are scared to take risks, and one of those risks is making decisions based on large amounts of data. Experts in information technology and data science have their work cut out for them when it comes to organizing data and gaining access to them. They have to come up with a system that is effective for

carrying out this task. So, if it is effectively handled and put to use, big data has the potential to be of great use to our existence, spanning from recognizing trends in the market to protecting oneself from illness and violence. When it comes to the care of patients, trust may be an even more important factor than access. Providers are unable to exploit any insights that may have been generated from unreliable, erroneous, or noisy data. This is the case even though it may be difficult to verify the accuracy of a dataset. According to an article that was published in the New York Times in 2014, data scientists often spend approximately 60% of their time cleaning the data so that they may be used. It is possible that this number is much greater for analysts who work in the healthcare industry. With so many systems allowing free text or other unstructured inputs, providers are always fighting to improve the integrity and quality of their data. Healthcare businesses must use data governance and information governance as critical measures to make sure that the data are accurate, full, standardized, and ready to use.

## 8.4  Big Data in Genomics

It is generally understood that genomics is the study of an organism's entire genetic makeup. Deoxyribonucleic acid (DNA), which has the structure of a double helix, is what makes up this genetic material. The chemical bases adenine, thymine, guanine, and cytosine make up the four main types of nucleotides that make up DNA. The ladder shape, commonly known as the double helix, is formed by twisting pairs of these four bases. Adenine and thymine, as well as guanine and cytosine, are the only pairs of bases that may exclusively link with one another. Technology geared toward digesting a genome is now required to usher in the era of personalized medicine. The complexity of biological systems necessitates the use of technology, which is necessary even to begin to comprehend the human genome because it would be practically impossible for human vision to comprehend around six million base pairs and identify every variation. In addition, new techniques are required to process the enormous volumes of data generated while searching a genome. This industry has transitioned into the world of big data as a result of the enormous quantity of data generated by the sequencing, mapping, and analysis of genomes. When one considers that each piece of genetic information has between 20,000 and 25,000 genes, which results in a total of six million base pairs per genome, one may get an idea of the amount of information that is included in each genome. This translates to 102,400 images or 100 terabytes of information per human genome [28]. It is believed that an individual living body holds more than 140 million petabytes of data [29]. Furthermore, infrastructure that operates quickly and effectively is required. The HGP historically required 13 years to complete, although many affected patients don't have to wait 13 years. Regrettably, in the realm of medicine, a mere 24 hours can make all the difference. The quicker a genome is sequenced, the quicker various illnesses and ailments can be discovered, and the quicker doctors can formulate a treatment strategy. Another element that needs to be taken into account is the cost of genome sequencing. When the initial steps in genome sequencing were taken 13 years ago, it cost $3 billion. Today, we are gradually getting closer to more affordable pricing. The cost has decreased from $300 million a decade ago to $3,000 now, and according to certain predictions, it will be less than $1,000 by the end of this year. Utilizing the promise of genomics and big data technologies will need a huge financial commitment and amount of work. The following is a list of several promising initiatives:

- *Gentle Labs*: A brand-new player in the personal genomics industry is called Gentle Labs. Gentle provides screenings for more than 1,700 genetic conditions, such as breast cancer, colon cancer, aneurysms, and heart problems. The company gives its customers a kit that includes a test tube to collect saliva for the screening, and the package also contains other necessary materials.
- The National Geographic Society and IBM collaborated to create the Genographic Project to map out the patterns of human migration by collecting and analyzing DNA samples from all across the world. Self-testing kits are currently available from the company. DNA is quickly extracted from the mouth, examined, and added to an internet database.
- On the flip side of the genomics issue is a firm called Genestack, which acts as "the Genomics Operating System" rather than studying DNA and patient genomes. Launched to use big data computing and application engineering to change genomics, they set out to lessen the inefficiencies now associated with using genomic tools and the widespread incompatibility of numerous products. Their operating system

is used to create, execute, and share large-scale genomic applications and datasets for pharma, biotech, and healthcare. Others include Positive Bioscience, 23andME, Expense Bioinformatics, Counsyl, and DNAnexus.

As the healthcare sector proceeds to undergo technological change, privacy is a subject of major concern. Since a user's genome may disclose their identities as well as those of their ancestors and divulge private details about their family background, data related to genomes are a topic of intense controversy. There is no longer a guarantee of privacy due to the vast volumes of data that are currently exchanged across different individuals and healthcare experts.

## 8.5 Architectural Framework

The conceptual framework of a traditional health informatics or analytical project is analogous to that of a study on big data analytics in the healthcare industry. The most important difference is based on how the processing is really carried out. Throughout the course of a typical health analytics project, an analysis may be carried out by making use of a business intelligence tool that has been installed on a standalone device, such as a desktop or laptop computer. Since big data, by definition, is extremely large, its processing must be partitioned and distributed across multiple nodes. The concept of distributed processing has been around for quite some time. It is very recent that it has been put to use in the analysis of extremely large datasets, as healthcare practitioners are beginning to make use of the massive data repositories at their disposal in order to gain knowledge and make decisions that are more beneficial to patients' health. In addition, cloud-based open-source technologies, such as Hadoop/ MapReduce, have helped popularize the application of big data analytics in the healthcare sector. The user interfaces of traditional tools and those used with big data are entirely different from one another, despite the fact that the methodology and models of both types of tools are comparable. The traditional care analytics tools that have been around for a while have been refined to become more open and user-friendly. On the other hand, big data analytics tools are incredibly complicated, involve a lot of programming, and call for the use of several abilities. They are largely open-source development tools and platforms that have been developed ad hoc. Hence, they lack vendor-driven proprietary products' support and user-friendliness. Big data in healthcare can originate from intrinsic (e.g., EHRs and CPOE) and exterior (official sources, research labs, drug stores, health insurers, health maintenance organizations, etc.) sources and is frequently present in a variety of genres (flat files, CSV files, relationship tables, ASCII/text, etc.) and areas (geographic, in addition to the locations of health professionals and patients) in countless lineages as well as other tools. These data must be gathered to do big data analytics. The second component involves processing or transforming the "raw" data, which has several possible outcomes. One option is to use web services (middleware) in conjunction with a service-oriented architectural approach [30]. Amenities are used to call, retrieve, and process the data, while the data remain unprocessed. Data warehousing is a different strategy in which data from many sources are combined and prepared for processing; however, the data are not immediately accessible. The extract, transform, and load (ETL) process cleans and prepares data from a variety of sources. The big data analytics platform may accept data in several different forms based on whether the data are organized or unstructured. Several choices are made in this next conceptual component of the framework involving the data input method, distributed design, tool choice, and analytics models.

The four common uses of big data analytics in healthcare are illustrated on the far right in Figure 8.2.

These consist of OLAP, data mining, reports, and queries. The four applications have a common theme of visualization. To gather, handle, analyze, and display big data in healthcare, a broad range of methodologies and technologies have been created and modified, drawing on various disciplines, including statistics, information science, numerical methods, and economics.

The open-source Hadoop (Apache platform), created for such mundane tasks as gathering online search indexes, is the most important platform for big data analytics. It belongs to the group of "NoSQL" systems that have developed to gather data in novel ways, together with CouchDB and MongoDB. Hadoop's ability to process incredibly huge volumes of data is mostly due to the allocation of partitioned datasets to several computers (nodes), where each one addresses a different aspect of the overall issue before integrating its findings to provide the desired outcome [31]. Hadoop may function as both an analytical tool and a data organizer. It has a lot of potential for helping businesses use data that have previously been challenging to manage and evaluate. These are undoubtedly serious

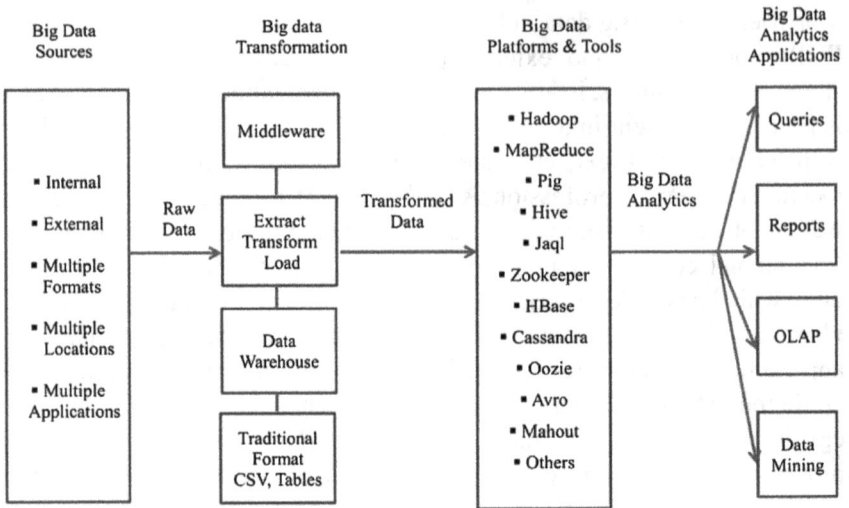

**Figure 8.2.** An Applied Architecture of Big Data Analytics

problems in the healthcare sector. Thus, trade-offs must be discussed. Additionally, these platforms/tools demand a lot of programming knowledge, which the common healthcare end user might not have. Furthermore, given that big data analytics in healthcare is a relatively new phenomenon, governance problems such as control, confidentiality, safety, and norms have not yet been resolved.

## 8.5.1 Methodology of Big Data Analytics in Healthcare

Despite the fact that many different ways are currently being developed for this rapidly expanding subject, we provide in this article one that is both helpful and interactive. In the first step, the multidisciplinary big data analytics team working on the healthcare project will write a "concept statement". This is the first step in attempting to demonstrate why such a project is absolutely necessary. After the initial statement of the idea comes an explanation of the importance of the endeavor. The healthcare practitioner will explain that there are some trade-offs that need to be made, including those regarding cost, scalability, and alternative options. The team may go on to Step 2, the proposal development stage, in which the concept statement will be accepted. Here, further information is provided. Several inquiries are answered in light of the concept statement: What issue is being dealt with? Why does the healthcare practitioner find it interesting and important? What are the benefits of a big data analytics strategy? (It is crucial to justify the usage of big data analytics because they are more expensive and sophisticated than traditional analytics methodologies). The project team should also include background data on the issue domain and examples of earlier work and studies in this area. The methodology's phases are then developed and put into practice in Step 3. Each proposition in the idea statement is split into separate sentences. Take note that these are not as thorough as statistical techniques would be. Instead, they are created to assist in directing the big data analytics process. The independent and dependent variables or indicators are recognized simultaneously. The data are gathered, documented, and converted in advance of analytics; the data sources, as shown in Figure 8.2, are also recognized. The examination and selection of platforms and tools is now a crucial stage. As previously mentioned, there are several choices, such as AWS Hadoop, Cloudera, and IBM BigInsights. The data will then be subjected to different big data analytics techniques. The only way this procedure varies from standard analytics is that massive datasets are used as a scale for the techniques. Insight is obtained from big data analytics through several iterations and what-if studies [32]. The understanding allows for the making of well-informed judgments. Step 4 involves testing and validating the models and presenting the results to the stakeholders for further action. With feedback loops built in at each level to reduce the likelihood of failure, implementation is done in stages.

## 8.5.2 Advantages of Big Data Analytics to Healthcare

Businesses within the healthcare industry stand to greatly benefit by properly digitizing, combining, and exploiting big data [33]. These businesses range from practices with a single physician and multi-provider groups to enormous hospital networks and accountable care organizations. The possible benefits include the ability to manage unique individual and group health, detect fraudulent healthcare swiftly and efficiently, and diagnose illnesses at an early point, so they can be managed more conveniently and proficiently. The analysis of large amounts of data can provide solutions to many problems. On the basis of enormous amounts of historical data, certain developments or outcomes can be predicted and/or estimated. These include the length of stay (LOS), clients who will vote for elective surgery, patients who are likely not to benefit from surgery, problems, people at risk for health complications, patients at high risk for sepsis, MRSA, C. difficile, or other care facility illnesses, ailment progression, patients at risk of serious illnesses, and reasons for the event (EMC consulting). Two of the sectors with the most room for reduction are clinical services and R&D, where waste amounts to $165 billion and $108 billion, respectively. Big data, according to McKinsey, might aid in reducing ineffectiveness and waste in the following key ways:

- *Clinical Operations*: Comparative effectiveness studies to find more affordable and clinically appropriate ways to assess and treat patients.
- *Public Health*: To enhance public health monitoring and hasten response, illness patterns are analyzed together with epidemics and transmission; faster creation of vaccines that are more precisely targeted, such as selecting the seasonal influenza strains, and the transformation of huge chunks of data into knowledge that can be utilized to establish priorities, offer services, and anticipate and preclude crises, particularly for the benefit of communities.

Furthermore, incident-based medicine, genetic analytics, surveillance systems, and client profile analysis may all benefit from big data analytics in the healthcare industry.

### 8.5.3 Challenges of Big Data in Healthcare

Big data has the potential to greatly benefit people, companies, nations, and the whole planet. Benefits, however, might also come with disadvantages, such as a loss of security and privacy. Furthermore, a lot of big data technologies are free to use and open source, which might lead to backdoors for invasions, hackers, and data theft. Therefore, it is important to evaluate reliability, secrecy, privacy, and authenticity. The assessment criteria for platforms may include accessibility, consistency, usability, scalability, and the capacity for manipulation at various levels of detail. They may also include quality control as well as privacy and security support. However, despite the fact that the vast majority of platforms currently in use are open source, the standard advantages and disadvantages of such platforms continue to apply. In order to be successful, big data analytics in the healthcare industry need to be presented in a way that is menu-driven, accessible, and fair [34].

The field of healthcare has a significant requirement for analytics of real-time big data. It is essential to find a solution to the time lag that exists between the collection of data and their processing. Adoption on a massive scale also necessitates the dynamic availability of a wide variety of analytics algorithms, models, and approaches organized in the format of a pull-down menu. It is imperative that the essential managerial concerns of ownership, governance, and standards be taken into consideration at all times. These issues, along with those of continual data gathering and data cleaning, are inextricably linked. Data pertaining to medical insurance are rarely standardized, frequently fragmented, or produced by outmoded information technology systems in formats that are not compatible [35]. It is essential to find a solution to this big issue as well.

## 8.6　Chapter Summary

In conclusion, the role of big data in healthcare is becoming increasingly important as the industry generates vast amounts of data. Big data analytics can provide valuable insights into patient care, disease prevention, and population health management. By analyzing large and complex datasets, healthcare organizations can identify patterns and trends that were previously difficult to detect, which can lead to more personalized and effective treatments.

However, there are also challenges associated with big data analytics in healthcare, such as data privacy and security concerns, the need for skilled data scientists and analysts, and the potential for bias in algorithms. To fully realize the potential of big data in healthcare, these challenges need to be addressed through appropriate regulations and training programs.

Overall, big data has the potential to transform healthcare by improving patient outcomes, reducing costs, and increasing efficiency. It will be exciting to see how healthcare organizations continue to harness the power of big data to drive innovation and advancements in patient care.

## References

1. Laney, D. (2001). *3D Data Management: Controlling Data Volume, Velocity, and Variety, Application Delivery Strategies*. Stamford: META Group Inc.
2. Shameer, K., Badgeley, M. A., Miotto, R., Glicksberg, B. S., Morgan, J. W., Dudley, J. T. (2017). Translational bioinformatics in the era of real-time

biomedical, health care and wellness data streams. *Brief Bioinformatics*, 18(1), 105–124.

3. Plageras, A. P., Stergiou, C., Kokkonis, G., Psannis, K. E., Ishibashi, Y., Kim, B. G., & Gupta, B. B. (2017). Efficient large-scale medical data (Ehealth Big Data) analytics in Internet of Things. In *2017 IEEE 19th Conference on Business informatics (CBI)*, 24–26 July 2017, Macau, China, (Vol. 2, pp. 21–27). IEEE. Thessaloniki, Greece.

4. Gupta, B. B., Gaurav, A., & Panigrahi, P. K. (2023). Analysis of security and privacy issues of information management Oo Big Data in B2B based healthcare systems. *Journal of Business Research*, 162, 113859.

5. Bansal, S., Chowell, G., Simonsen, L., Vespignani, A., & Viboud, C. (2016). Big data for infectious disease surveillance and modeling. *The Journal of Infectious Diseases*, 214, S375–S379.

6. Bates, D. W., Saria, S., Ohno-Machado, L., Shah, A., & Escobar, G. (2014). Big data in health care: Using analytics to identify and manage high-risk and high-cost patients. *Health Affairs*, 33, 1123–1131.

7. Khoury, M. J. & Ioannidis, J. P. A. (2014). Big data meets public health. *Science*, 346, 1054–1055.

8. Kostkova, P. (2013). A roadmap to integrated digital public health surveillance: The vision and the challenges. In *Proceedings of the International Conference on World Wide Web, 22nd, Rio de Janeiro*, 13–17 May 2013, Rio de Janeiro, Brazil, (pp. 687–694). New York: Assoc. Comput. Mach. (ACM)

9. Teutsch, S. M. & Churchill, R. E. (2000). *Principles and Practice of Public Health Surveillance*. Oxford, UK: Oxford University Press.

10. Casillo, M., Colace, F., Gupta, B. B., Marongiu, F., & Santaniello, D. (2021). Decentralized Approach for Data Security of Medical IoT Devices. In *International Conference on Smart Systems and Advanced Computing*, 30–31 December 2022, *(Syscom-2021)*, Macau, China.

11. Gomide, J., Veloso, A., Meira, W. Jr., Almeida, V., Benevenuto, F., Ferraz, F., & Teixeria, M. (2011). Dengue surveillance based on a computational model of spatio-temporal locality of Twitter. In *Proceedings of the 3rd International Web Science Conference on Koblenz, Germany*, Art. 3. Assoc. Comput. Mach. (ACM), June 15–17, 2011, Koblenz, Germany.

12. Anderson, T. K. (2009). Kernel density estimation and $K$-means clustering to profile road accident hotspots. *Accident Analysis and Prevention*, 41, 359–364.

13. Holmes, E., Loo, R. L., Stamler, J., Bictash, M., Yap, I. K. S., Chan, Q., Ebbels, T., Iorio, M. D., Brown, I. J., Veselkov, K. A., Daviglus, M. L., Kesteloot, H., Ueshima, H., Zhao, L., Nicholson, J. K., & Elliott, P. (2008). Human metabolic phenotype diversity and its association with diet and blood pressure. *Nature*, 453, 396–400.

14. Fahmi, P., Viet, V., & Deok-Jai, C. (2012). Semi-supervised fall detection algorithm using fall indicators in smartphone. In *Proceedings of the 6th International Conference on Ubiquitous Information Management and Communication, Kuala Lumpur, Malaysia*, Art. 122. Assoc. Comput. Mach. (ACM), New York.
15. Davis, H. T., Aelion, C. M., McDermott, S., & Lawson, A. B. (2009). Identifying natural and anthropogenic sources of metals in urban and rural soils using GIS-based data, PCA, and spatial interpolation. *Environmental Pollution*, 157, 2378–2385.
16. Egger, M. E., Squires, M. H., Kooby, D. A., Maithel, S. K., Cho, C. S., Weber, S. M., Winslow, E. R., Martin, R, C, G., McMasters, K. M., Scoggins, C. R. (2015). Risk stratification for readmission after major hepatectomy: Development of a readmission risk score. *Journal of the American College of Surgeons*, 220, 640–648.
17. Mamiya, H., Schwartzman, K., Verma, A., Jauvin, C., Behr, M., & Buckeridge, D. (2015). Towards probabilistic decision support in public health practice: Predicting recent transmission of tuberculosis from patient attributes. *Journal of Biomedical Informatics*, 53, 237–242.
18. Kononen, D. W., Flannagan, C. A., & Wang, S. C. (2011). Identification and validation of a logistic regression model for predicting serious injuries associated with motor vehicle crashes. *Accident Analysis and Prevention*, 43, 112–122.
19. De Choudhury, M., Kiciman, E., Dredze, M., Coppersmith, G., & Kumar, M. (2016). Discovering shifts to suicidal ideation from mental health content in social media. In *Proceedings of the 2016 CHI Conference on Human Factors in Computing Systems, San Jose, CA* (pp. 2098–2110). Assoc. Comput. Mach. (ACM), New York.
20. Tilson, H. & Gebbie, K. M. (2004). The public health workforce. *Annual Review of Public Health*, 25, 341–356.
21. Wing, J. M. (2006). Computational thinking. *Communications of the ACM*, 49, 33–35.
22. Kochenderfer, M. J. (2015). *Decision Making Under Uncertainty: Theory and Application*. MIT Press, Cambridge, MA.
23. Gaurav, A. & Chui, K. T. (2022). Advancement of cloud computing and big data analytics in healthcare sector security. *Data Science Insights Magazine*, (Vol. 1, pp. 12–15), Insights2Techinfo.
24. Lichtveld, M. Y. (2016). A Timely Reflection on the Public Health Workforce. *Journal of Public Health Management & Practice*, 22, 509–511.
25. Mehla, R. (2021). "Application of Deep Learning in Big Data Analytics for Healthcare Systems", Insights2Techinfo, p. 1. https://insights2techinfo.com/application-of-deep-learning-in-big-data-analytics-for-healthcare-systems/.

26. Spiegelman, D. (2016). Evaluating public health interventions: 2. Stepping up to routine public health evaluation with the stepped wedge design. *American Journal of Public Health*, 106, 453–457.
27. Hafeman, D. M. & Schwartz, S. (2009). Opening the Black Box: A motivation for the assessment of mediation. *International Journal of Epidemiology*, 38, 838–845.
28. Feldman, B., Martin, E. M., & Skotnes, T. (2013). Genomics and the role of big data in personalizing the healthcare experience. O'Reilly Data.
29. Shrestha, R. B. (2014). Big data and cloud computing. *Applied Radiology*, 43(3), 32–35.
30. Raghupathi, W. & Kesh, S. (2007). Interoperable electronic health records design: Towards a service-oriented architecture. *e-Service Journal*, 5, 39–57.
31. Zikopoulos, P., Deroos, D., Parasuraman, K., Deutsch, T., Giles, J., & Corrigan, D. (2012). Harness the power of big data The IBM big data platform. McGraw Hill Professional, New York, U.S.A.
32. Mamta (2021). "Quick Medical Data Access Using Edge Computing", Insights2Techinfo, p. 1. https://insights2techinfo.com/quick-medical-data-access-using-edge-computing/.
33. Burghard, C. (2012). Big data and analytics key to accountable care success. *IDC Health Insights*, 1, 1–9.
34. Gaurav, A., Psannis, K., & Peraković, D. (2022). Security of cloud-based medical internet of things (miots): A survey. *International Journal of Software Science and Computational Intelligence (IJSSCI)*, 14(1), 1–16.
35. IHTT (2013). "Transforming Health Care through Big Data Strategies for leveraging Big Data in the Health Care Industry". http://ihealthtran.com/wordpress/2013/03/iht%C2%B2-releases-big-data-research-report download-today/.

# Chapter 9

# Big Data in Finance

*Big data* refers to the vast amount of information generated by individuals, organizations, and machines in our modern digital world. The financial industry is one of the areas that generates a large amount of data, including financial transactions, stock prices, economic indicators, and other financial market data [1]. The integration of big data into financial research has the potential to provide insights and predictions that can inform investment decisions, market analysis, and risk management [2]. The financial industry has been quick to embrace big data, and the use of big data in financial research has increased significantly in recent years. However, the use of big data in financial research is not without its challenges. The sheer volume of data can make it difficult to extract meaningful insights, and the accuracy and reliability of the data must be carefully evaluated. This chapter explores the application of big data in financial research, including the sources of financial data, the types of analysis used in financial research, and the challenges associated with using big data in financial research.

## 9.1 Digitalization in Financial Industry

Digitalization has had a profound impact on the financial industry, transforming the way financial institutions operate and deliver services to customers [3]. Some of the key ways digitalization has impacted the financial industry include:

- *Online Banking*: With the advent of digitalization, online banking has become ubiquitous. Customers can now access their bank accounts, transfer funds, pay bills, and check their account balances from anywhere at any time.
- *Mobile Banking*: Mobile banking has become increasingly popular, allowing customers to access banking services through their smartphones and other mobile devices.
- *Electronic Payments*: Electronic payment systems, such as credit cards, debit cards, and digital wallets, have become the norm, making transactions faster and more convenient.
- *Automated Trading*: Digitalization has also transformed trading in financial markets. Automated trading algorithms now execute trades at lightning-fast speeds based on complex mathematical models.
- *Data Analysis*: Financial institutions now have access to vast amounts of data, which they can use to analyze customer behavior, develop new products, and make more informed decisions.
- *Blockchain Technology*: Blockchain technology has disrupted the financial industry, enabling secure, transparent, and efficient transactions without the need for intermediaries.
- *Robo-Advisory*: Robo-advisory services, which use algorithms to provide investment advice, have become increasingly popular, providing customers with low-cost investment options.

Overall, digitalization has transformed the financial industry, providing customers with greater convenience, faster transactions, and more personalized services while enabling financial institutions to operate more efficiently and effectively [4]. Digitalization has played a key role in enabling the financial industry to harness the power of big data. With the advent of digital technologies, financial institutions now have access to vast amounts of data, including customer transaction data, social media data, and other sources of information [5]. These data can be analyzed using advanced analytics tools to gain insights into customer behavior, identify patterns and trends, and make more informed decisions.

Some of the key ways in which digitalization and big data have impacted the financial industry include:

- *Improved Risk Management*: Financial institutions can use big data analytics to better understand and manage risks associated with lending and investments. This includes identifying potential risks and opportunities, as well as monitoring and predicting market trends.

- *Personalized Services*: By analyzing customer data, financial institutions can provide more personalized services tailored to individual customer needs and preferences. This can help improve customer satisfaction and loyalty.
- *Fraud Detection*: Big data analytics can help financial institutions identify and prevent fraud by analyzing patterns of behavior and transaction data to identify suspicious activity.
- *Improved Compliance*: With access to large amounts of data, financial institutions can more easily meet regulatory compliance requirements, including reporting and monitoring requirements.
- *Enhanced Decision-Making*: By leveraging big data analytics, financial institutions can make more informed decisions based on data-driven insights. This can help improve business performance and profitability.

Thus, big data has a transformative impact on the finance industry, enabling financial institutions to make more informed decisions, improve risk management, and provide more personalized services to their customers.

## 9.2 Sources of Financial Data

There are many sources of financial data, and each source has its own advantages and disadvantages. Some of the most common sources of financial data include [6]:

- *Financial Transactions*: This includes data from credit card transactions, bank transfers, and other financial transactions. Financial transaction data are highly detailed and can provide insights into consumer spending patterns and financial behavior.
- *Stock Prices*: This includes data on stock prices, dividends, and other stock market information. Stock prices are a key indicator of a company's financial performance, and the analysis of stock prices can provide valuable insights into the health of the overall stock market.

- *Economic Indicators*: This includes data on GDP, inflation, unemployment rates, and other economic indicators. Economic indicators can provide valuable insights into the health of the economy and the financial industry.
- *News and Social Media*: This includes data from news articles, social media posts, and other sources of information that can provide insights into market sentiment and consumer behavior.
- *Regulatory Data*: This includes data from regulatory agencies, such as the Securities and Exchange Commission (SEC) and the Financial Industry Regulatory Authority (FINRA). Regulatory data provide valuable information on a company's financial performance, including its financial statements, earnings reports, and other financial disclosures.

The data from these sources come in large volume, velocity, and variety, indicating that their nature matches the characteristics of big data. Volume here refers to the vast amounts of data generated by financial transactions, market data, and other sources. Financial institutions are now able to collect and analyze huge volumes of data from a variety of sources, including structured and unstructured data, to gain insights into customer behavior, risk management, and trading strategies [7]. Velocity refers to the speed at which data are generated, collected, and analyzed. Financial institutions need to be able to process and analyze data in real time in order to make informed decisions quickly. With the increasing use of high-speed trading algorithms and other real-time analytics tools, velocity has become an increasingly important factor in financial data analysis. Finally, variety refers to the diversity of data types and sources, including texts, images, videos, social media, and other sources. Financial institutions need to be able to collect, analyze, and integrate data from a wide range of sources in order to gain a comprehensive view of customer behavior and market trends. With the increasing use of alternative data sources, such as social media sentiment analysis and satellite imagery, variety has become an increasingly important factor in financial data analysis. In order to effectively manage the three V's of big data in finance, financial institutions need to invest in robust data management systems, powerful analytics tools, and skilled data scientists and analysts. By doing so, they can leverage the power of big data to gain insights into customer behavior, improve risk management, and make more informed decisions.

## 9.3 Challenges of Using Big Data in Financial Research

Despite the many benefits of using big data in financial research, there are also many challenges associated with this approach. Some of the key challenges include:

- *Data Quality*: Financial data can be highly complex and difficult to clean and validate. The accuracy and reliability of the data must be carefully evaluated to ensure that they are suitable for analysis.
- *Data Security*: Financial data are highly sensitive and must be protected from unauthorized access or disclosure. The use of big data in financial research raises concerns about data privacy and security.
- *Data Integration*: Financial data are often siloed in different systems and formats, which can make it difficult to integrate and analyze. Integration challenges can arise due to differences in data formats, protocols, and structures.
- *Bias*: The analysis of financial data can be subject to bias, both in terms of the data itself and the analysis techniques used. The use of biased data or analysis techniques can lead to inaccurate or misleading results.
- *Scalability*: The volume of financial data can be overwhelming, making it difficult to scale up analysis techniques to handle large datasets.
- *Regulatory Compliance*: The use of big data in financial research must comply with a range of regulatory requirements, including data protection laws, anti-money laundering regulations, and other financial regulations.
- *Interpreting Results*: The analysis of big data can produce large amounts of results, which can be difficult to interpret and act upon. It is important to have the right tools and expertise to turn the analysis into actionable insights.

The challenges associated with using big data in financial research must be carefully considered. Addressing these challenges requires a combination of technical expertise, data governance, and regulatory compliance measures. With the right approach, big data can be a powerful tool for driving innovation and improving the financial industry.

## 9.4  Financial Big Data

Financial big data (FBD) refers to the vast amount of financial data generated through various sources, such as stock exchanges, trading platforms, news outlets, social media, and other financial institutions [8]. The data are usually diverse in terms of type, volume, velocity, and veracity, and require specialized tools and technologies to process and analyze them. FBD analysis involves applying advanced analytical techniques, such as machine learning, data mining, and predictive modeling, to extract meaningful insights and patterns from large datasets. These insights can be used for various purposes such as risk management, fraud detection, investment decisions, and improving customer experience. FBD has become increasingly important in the financial industry, as it enables financial institutions to make more informed decisions based on data-driven insights [9]. The rise of technologies such as blockchain, cloud computing, and artificial intelligence has also contributed to the growth of financial big data management and analytics.

### 9.4.1  FBD Management

FBD management refers to the process of organizing, storing, analyzing, and utilizing large volumes of financial data. Effective management of FBD can provide valuable insights into consumer behavior, market trends, and financial risks, enabling better investment decisions, risk management, and compliance monitoring [10].

Here are some of the key aspects of FBD management:

- *Data Governance*: Data governance refers to the process of managing the availability, usability, integrity, and security of data used in an organization. In the context of financial big data management, data governance ensures that financial data are accurate, reliable, and secure. It involves establishing policies and procedures for data management, defining data quality standards, and monitoring compliance with regulatory requirements.
- *Data Storage and Processing*: FBD requires a robust and scalable infrastructure for storage and processing. This typically involves the use of cloud-based or on-premises data storage solutions, such as Hadoop or Apache Spark. These solutions allow for the storage and processing of large volumes of data in a distributed and parallel manner, enabling faster and more efficient data analyses.

- *Data Integration*: FBD often comes from multiple sources, such as financial institutions, regulatory bodies, and third-party vendors. Integrating data from these sources requires careful consideration of data formats, data protocols, and data quality. Data integration involves the use of data integration tools and techniques, such as extract, transform, and load (ETL) processes, to ensure that data are consistent and accurate.
- *Data Analysis*: Data analysis is a crucial aspect of FBD management. It involves the use of statistical, machine learning, and other analytical techniques to extract insights from large volumes of data. The analysis of FBD can provide valuable insights into consumer behavior, market trends, and financial risks.
- *Data Visualization*: Data visualization is an important component of FBD management. It involves the use of charts, graphs, and other visual aids to present data in a meaningful and accessible way. Effective data visualization can help communicate complex financial data to a wider audience, enabling better decision-making.
- *Regulatory Compliance*: FBD management must comply with a range of regulatory requirements, such as data protection laws, anti-money laundering regulations, and other financial regulations. Compliance with these regulations requires careful consideration of data privacy, security, and retention policies.

Effective FBD management is critical to the success of financial institutions. By effectively managing FBD, financial institutions can gain valuable insights into consumer behavior, market trends, and financial risks, enabling better investment decisions, risk management, and compliance monitoring.

### 9.4.2 FBD Analytics

The use of big data in financial research requires the application of various data analysis techniques [11]. Some of the most common types of analysis used in financial research include:

- *Statistical Analysis*: This includes the use of statistical methods to analyze financial data, including regression analysis, time-series analysis, and hypothesis testing. Statistical analysis is a powerful tool for identifying patterns and trends in financial data.

- *Machine Learning*: This includes the use of algorithms to analyze financial data and make predictions. Machine learning algorithms can be used to predict stock prices, identify trends in financial data, and analyze consumer behavior.
- *Natural Language Processing (NLP)*: This includes the use of computational methods to analyze text data, including news articles and social media posts. NLP can be used to analyze sentiment and identify key themes and topics in financial data.
- *Network Analysis*: This includes the use of graph theory and network analysis to study the relationships between different financial entities, including stocks, investors, and companies. Network analysis can provide valuable insights into the relationships between different financial entities and help identify potential risks and opportunities.

The key steps involved in implementing big data analytics in finance are as follows [12]:

- *Identify Use Cases*: The first step in implementing big data analytics in finance is to identify the use cases where big data analytics can provide the most value. This can include areas such as risk management, fraud detection, customer segmentation, and marketing.
- *Data Collection and Integration*: Once the use cases have been identified, the next step is to collect and integrate relevant data from multiple sources. This can include transaction data, customer data, market data, and external data sources, such as social media and news feeds.
- *Data Cleaning and Preparation*: The quality of the data is critical to the success of big data analytics. Data cleaning and preparation involve removing duplicates, correcting errors, and standardizing data formats to ensure that the data are accurate, complete, and consistent.
- *Data Analysis*: The data are then analyzed using a range of analytical techniques, such as machine learning, predictive analytics, and data mining. This can include identifying patterns, detecting anomalies, and making predictions.
- *Visualization and Reporting*: The results of the analysis are then visualized and reported in a way that is easy to understand and actionable. This can include dashboards, reports, and alerts that highlight the key insights and trends.

- *Integration with Business Processes*: Big data analytics should be integrated with the business processes of the financial institution to ensure that the insights generated are used to drive better decision-making. This can include integrating the results of the analysis into risk management processes, customer segmentation strategies, and marketing campaigns.
- *Regulatory Compliance*: Finally, it is important to ensure that the implementation of big data analytics in finance complies with regulatory requirements, such as data protection laws, anti-money laundering regulations, and other financial regulations.

By following these steps, financial institutions can effectively leverage the power of big data analytics to drive better decision-making and gain a competitive advantage in the market.

## 9.5 Theoretical Framework of Big Data in Financial Services

The theoretical framework of big data in financial services refers to the conceptual foundation upon which the use of big data in finance is based. In the majority of instances, big data in finance is associated with four types of financial industries, namely the financial market, online marketplace, lending organization, and bank, as shown in Figure 9.1 [13]. These industries generate silos of data each day as a result of their daily transactions, user accounts, data updates, and account modifications.

As the data are coming in large volumes from heterogeneous resources with high velocity, all these constitute the key characteristics of big data. Therefore, it requires careful data integration and parallel processing in order to extract valuable insights for actional decision-making that can be used to identify frauds, predict risks, forecast pricing, etc.

## 9.6 Popular Use Cases of FBD Analytics

There are numerous methods for financial institutions to improve their offerings, cut expenses, and increase client satisfaction through the implementation of big data analytics. Financial big data analytics has several use cases in the financial services industry, which include banking,

Social Media
BANK
Lending Company
Online Marketplace
Financial Market

Verification & Collection | Credit Risk Prediction | Fraud Detection

Return Prediction
Volatility Forecast
Market Valuation
Excess Trading Volume
Risk Analysis
Portfolio Management
Co-Movement
Option Pricing
Algorithmic Trading
Idiosyncratic Volatility

Heterogeneous Data Sources
Identity Matching & Data integration
User Account
Variables/Data Fields
Financial Services
E-Commerece Activities
Others Activities
Capital Market

Financial Big Data
Stock Message Board, Whisper Forecasts, Search Engines, Spam

Conventional Media
Newspapers, Advertising, Radio, Television...

Stock Market Data
Price, Volume, Transaction Cost, Short Interest.

Role of Financial Market Data

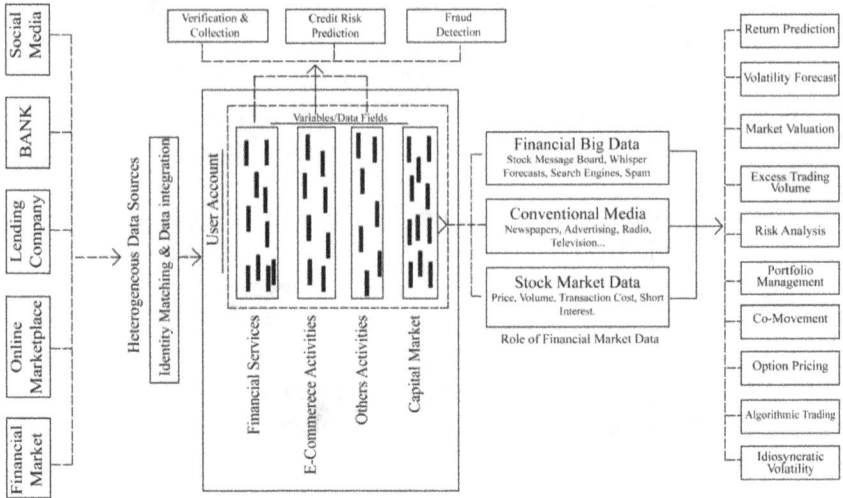

**Figure 9.1.**   Theoretical Framework of Big Data in Financial Services [13]

insurance, investment management, and other related fields [14]. Here are some of the most common use cases of financial big data analytics:

- *Fraud Detection*: Financial institutions use big data analytics to detect and prevent fraud by analyzing large volumes of data to identify patterns and anomalies that indicate fraudulent activities [15]. This includes transactional data, customer behavior, and external data sources, such as social media and public records. For example, a bank can monitor consumer spending habits in order to discover any anomalous real-time financial activities. Big red flags will be raised if a normally frugal spender suddenly begins taking out loans and going on buying sprees.
- *Risk Management*: Financial institutions use big data analytics to identify and mitigate risks associated with financial transactions and investments. This includes analyzing market data, credit risk, operational risk, and other factors that impact the performance of financial products [16].

All financial institutions generate profits by making profitable investments. This may involve offering loans to individuals who will repay them or selling auto insurance to safe drivers. This was formerly the responsibility of specialists who would make educated guesses about which investments would likely be profitable. Now, the financial institutions evaluate massive databases of information to determine which investments are likely to be profitable and which are not, and they do so with astounding precision. This is utilized by banks to calculate the risk of various homebuyers and establish interest rates.

- *Customer Segmentation*: Financial institutions use big data analytics to segment customers based on their behavior, preferences, and other factors. This helps financial institutions personalize products and services to meet the specific needs of each customer segment [17].

  Banks can utilize data on consumer purchases to determine which goods, mortgages, and credit cards various categories of customers are likely to purchase. Companies use this information to segment customers into groups with comparable financial objectives and to pinpoint their advertising and sales efforts. Customers receive more recommendations for things they are likely to want, while banks spend less time attempting to market to people who are unlikely to purchase.

- *Marketing and Sales*: Financial institutions use big data analytics to identify potential customers and target them with relevant products and services. This includes analyzing customer data, social media data, and other external data sources to identify potential customers and develop targeted marketing campaigns [18,19].

  One way to utilize social media data is by using sentiment analysis, or opinion mining. This entails employing sophisticated NLP algorithms to determine what people truly believe based on what they say. These predictive analytics approaches are used to analyze millions of tweets and Facebook comments in order to make educated guesses about what is occurring in the world at this moment and to take immediate action. Numerous businesses use this to develop trading algorithms that can instantly bet against stock prices after a natural disaster. If you are the first person to learn that a plane has crashed, you can quickly bet that SpiceJet's stock price will fall, resulting in a substantial profit.

- *Investment Analysis*: Investment firms use big data analytics to analyze market trends, identify opportunities, and make informed investment decisions. This includes analyzing market data, economic indicators, and other factors that impact the performance of different asset classes [20].

- *Credit Scoring*: Financial institutions use big data analytics to assess the creditworthiness of individuals and businesses. This includes analyzing credit history, payment behavior, and other factors that impact creditworthiness [21,22].

- *Compliance and Risk Monitoring*: Financial institutions use big data analytics to monitor and enforce regulatory compliance and mitigate risk. This includes monitoring transactions, identifying potential violations, and developing strategies to prevent regulatory breaches [23,24].

FBD analytics has numerous use cases in the financial services industry, including fraud detection, risk management, customer segmentation, marketing and sales, investment analysis, credit scoring, and compliance and risk monitoring. By leveraging the power of big data analytics, financial institutions can improve their decision-making, enhance customer experience, increase operational efficiency, and mitigate risks.

## 9.7 Chapter Summary

In conclusion, big data has become an increasingly important tool in the financial services industry. Financial institutions are leveraging big data to improve decision-making, increase operational efficiency, enhance customer experience, and mitigate risks. By analyzing large volumes of data, financial institutions can identify patterns and trends that provide valuable insights into customer behavior, market trends, and other factors that impact financial performance.

The use of big data in finance has its challenges, including data management, data quality, privacy concerns, and regulatory compliance. However, financial institutions can overcome these challenges by implementing effective data management strategies, leveraging advanced analytics tools, and ensuring compliance with relevant regulations.

Overall, the use of big data in finance is expected to continue to grow as financial institutions seek to remain competitive in an increasingly data-driven industry. By effectively managing and leveraging big data, financial institutions can gain a competitive advantage and provide better products and services to their customers.

# References

1. Chaklader, B., Gupta, B. B., & Panigrahi, P. K. (2023). Analyzing the progress of FINTECH-companies and their integration with new technologies for innovation and entrepreneurship. *Journal of Business Research*, 161, 113847.
2. Goldstein, I., Spatt, C. S., & Ye, M. (2021). Big data in finance. *The Review of Financial Studies*, 34(7), 3213–3225.
3. Adrian, T. & Mancini-Griffoli, T. (2021). The rise of digital money. *Annual Review of Financial Economics*, 13, 57–77.
4. Chaudhary, P., Gupta, B. B., Chang, X., Nedjah, N., & Chui, K. T. (2021). Enhancing big data security through integrating XSS scanner into fog nodes for SMEs gain. *Technological Forecasting and Social Change*, 168, 120754.
5. Vives, X. (2019). Digital disruption in banking. *Annual Review of Financial Economics*, 11, 243–272.
6. Koop, G. (2022). *Analysis of Financial Data*. John Wiley & Sons Inc., Hoboken, New Jersey, USA.
7. Stergiou, C., Psannis, K. E., Gupta, B. B., & Ishibashi, Y. (2018). Security, privacy & efficiency of sustainable cloud computing for big data & IoT. *Sustainable Computing: Informatics and Systems*, 19, 174–184.
8. Trelewicz, J. Q. (2017). Big data and big money: The role of data in the financial sector. *IT Professional*, 19(3), 8–10.
9. Sharma, A., Singh, S. K., Badwal, E., Kumar, S., Gupta, B. B., Arya, V., Chui K. T., & Santaniello, D. (2023). Fuzzy based clustering of consumers' Big Data in industrial applications. In *2023 IEEE International Conference on Consumer Electronics (ICCE)* (pp. 1–3), 6–8 January 2023, Las Vegas, NV, USA. IEEE.
10. Munar, A., Chiner, E., & Sales, I. (2014). A big data financial information management architecture for global banking. In *2014 International Conference on Future Internet of Things and Cloud* (pp. 385–388). IEEE.
11. Cao, M., Chychyla, R., & Stewart, T. (2015). Big data analytics in financial statement audits. *Accounting Horizons*, 29(2), 423–429.
12. Hajiheydari, N., Delgosha, M. S., Wang, Y., & Olya, H. (2021). Exploring the paths to big data analytics implementation success in banking and financial service: An integrated approach. *Industrial Management & Data Systems*, 121(12), 2498–2529.
13. Hasan, M., Popp, J., & Oláh, J. (2020). Current landscape and influence of big data on finance. *Journal of Big Data*, 7(1), 1–17.
14. Srivastava, U. & Gopalkrishnan, S. (2015). Impact of big data analytics on banking sector: Learning for Indian banks. *Procedia Computer Science*, 50, 643–652.

15. Tang, J. & Karim, K. E. (2019). Financial fraud detection and big data analytics–implications on auditors' use of fraud brainstorming session. *Managerial Auditing Journal*, 34(3), 324–337.
16. Dicuonzo, G., Galeone, G., Zappimbulso, E., & Dell'Atti, V. (2019). Risk management 4.0: The role of big data analytics in the bank sector. *International Journal of Economics and Financial Issues*, 9(6), 40.
17. Chang, M. S. & Kim, H. J. (2018). A customer segmentation scheme base on big data in a bank. *Journal of Digital Contents Society*, 19(1), 85–91.
18. Rozados, I. V. & Tjahjono, B. (2014). Big data analytics in supply chain management: Trends and related research. In *6th International Conference on Operations and Supply Chain Management* (Vol. 1, p. 13), 10–13 December 2014, Bali, Indonesia.
19. Mohanty, S., Jagadeesh, M., & Srivatsa, H. (2013). *Big Data Imperatives: Enterprise 'Big Data' warehouse, 'BI' Implementations and Analytics.* Apress, Springer Nature, Switzerland AG.
20. Sun, Y., Shi, Y., & Zhang, Z. (2019). Finance big data: Management, analysis, and applications. *International Journal of Electronic Commerce*, 23(1), 9–11.
21. Onay, C. & Öztürk, E. (2018). A review of credit scoring research in the age of Big Data. *Journal of Financial Regulation and Compliance*, 26(3), 382–405.
22. Óskarsdóttir, M., Bravo, C., Sarraute, C., Vanthienen, J., & Baesens, B. (2019). The value of big data for credit scoring: Enhancing financial inclusion using mobile phone data and social network analytics. *Applied Soft Computing*, 74, 26–39.
23. Cheng, X., Liu, S., Sun, X., Wang, Z., Zhou, H., Shao, Y., & Shen, H. (2021). Combating emerging financial risks in the big data era: A perspective review. *Fundamental Research*, 1(5), 595–606.
24. Bendre, M. R. & Thool, V. R. (2016). Analytics, challenges and applications in big data environment: A survey. *Journal of Management Analytics*, 3(3), 206–239.

# Chapter 10

# Enabling Tools and Technologies for Big Data Analytics

The field of big data analytics relies on a wide range of enabling technologies and tools to store, process, and analyze large volumes of data. Some of the most important tools and technologies for big data analytics are as follows [1]:

- *Data Warehousing*: A data warehouse is a centralized repository that stores large volumes of structured and unstructured data. Data warehouses are designed to support complex analytical queries and provide fast, efficient access to data. Some popular data warehousing technologies include Amazon Redshift, Google BigQuery, and Microsoft Azure SQL Data Warehouse.
- *Hadoop*: Apache Hadoop is an open-source software framework that provides a distributed storage and processing system for large datasets. Hadoop is designed to handle both structured and unstructured data and can scale to accommodate massive amounts of data. Hadoop includes several key components, including the Hadoop Distributed File System (HDFS) and MapReduce for distributed processing.
- *NoSQL Databases*: NoSQL databases are designed to handle large volumes of unstructured or semi-structured data. These databases are typically faster and more scalable than traditional relational databases and can handle a wide range of data types. Popular NoSQL databases include Apache Cassandra, MongoDB, and Couchbase.

- *Data Integration*: Data integration tools are used to combine data from multiple sources and prepare it for analysis. These tools can handle complex data transformations and ensure data quality and consistency. Popular data integration tools include Apache NiFi, Talend, and Informatica.
- *Data Visualization*: Data visualization tools are used to create interactive, visual representations of data. These tools can help analysts and business users better understand and communicate insights from large datasets. Popular data visualization tools include Tableau, Power BI, and QlikView.
- *Machine Learning*: Machine learning algorithms are used to analyze large datasets and identify patterns and insights. These algorithms can learn from data and improve their accuracy over time. Popular machine learning libraries include TensorFlow, Scikit-Learn, and Apache Spark MLlib.
- *Cloud Computing*: Cloud computing platforms provide on-demand access to computing resources, including storage, processing, and analytics tools. Cloud platforms are highly scalable and can be used to process and analyze massive amounts of data. Popular cloud platforms for big data analytics include Amazon Web Services (AWS), Microsoft Azure, and Google Cloud Platform.

Overall, these enabling tools and technologies provide the infrastructure and capabilities needed to store, process, and analyze large volumes of data. By leveraging these tools, organizations can gain valuable insights and drive better decision-making in a wide range of industries, including finance.

## 10.1 Big Data Management and Modeling Tools

Cloudera Virtual Machine (VM) is a pre-configured virtual machine image that includes Cloudera's Hadoop distribution, along with a variety of other tools and software for big data processing and analytics [2]. It is designed to run on a virtualization platform, such as Oracle VirtualBox or VMware. The Cloudera VM provides a convenient way for developers, data analysts, and other users to get up and running with Hadoop quickly and easily without the need for complex setup and configuration. It includes a variety of Hadoop-related tools and software, including Apache

Hadoop, HDFS, Yet Another Resource Negotiator (YARN), MapReduce, Pig, Hive, Impala, Sqoop, and many others. Before one starts using various tools for big data, it is suggested to have Cloudera VM in place.

### 10.1.1 Data Modeling Tools

Data modeling is the process of creating a conceptual representation of data objects and their relationships with each other. There are various data modeling tools available on the market that can help organizations design, visualize, and document their data models efficiently [3]. The following sections discuss two popular tools for modeling vector and graph data.

### 10.1.2 Vector Data Model with Lucene

The vector data model is a method of representing spatial data in a geographic information system (GIS). It represents geographic features as points, lines, and polygons using $x$–$y$ coordinate on a two-dimensional plane. In this model, each feature is described by its geometry (shape and size) and its attributes (information about the feature, such as its name, type, and characteristics) [4,5].

For example, a road can be represented as a line in the vector data model, with attributes such as its name, speed limit, and number of lanes. A forest can be represented as a polygon with attributes such as its type, age, and size. The vector data model is commonly used in GIS applications because it allows for more precise measurements and analysis of geographic features. It is also easier to edit and update than other data models, such as the raster data model.

Lucene is primarily a search engine library that provides powerful text search capabilities. However, it also provides support for working with vector data in the form of vector space models. In Lucene, a vector space model is represented using a combination of document vectors and query vectors. Document vectors represent the content of documents in the collection, while query vectors represent the user's search query. The similarity between document vectors and query vectors is used to rank the documents in the collection based on their relevance to the user's search.

To work with vector data in Lucene, you can use the VectorField class, which represents a field in a Lucene document that contains vector data. A VectorField can be constructed using a SparseVector or a DenseVector, which are both implementations of the Vector interface.

Here's an example of how to create a document vector using a SparseVector in Lucene.

```
SparseVector vector = new SparseVector(3);
vector.set(0, 1.0);
vector.set(2, 2.0);
Document doc = new Document();
doc.add(new VectorField("myfield", vector, VectorField.TYPE.
SPARSE));
```

In this example, we create a SparseVector with three dimensions and set the values of the first and third dimensions to 1.0 and 2.0, respectively. We then create a new document and add a VectorField to it with the name "myfield" and the SparseVector we just created. Once you have created document vectors and query vectors, you can use Lucene's search capabilities to find documents that are similar to the query vector. Lucene provides several different similarity measures that can be used to compare document vectors and query vectors, including the cosine similarity measure.

Overall, Lucene's support for vector data models makes it a powerful tool for working with collections of documents that contain vector data. By representing document vectors and query vectors as fields in Lucene documents, you can easily perform a similarity-based search on the collection to find the most relevant documents to a given query.

### 10.1.3  Graph Data Model with Gephi

The graph data model is a method of representing data as a set of nodes and edges, also known as vertices and edges. It is a mathematical concept that is often used in computer science and information technology to model relationships between entities. In the graph data model, nodes represent entities, such as people, places, or things, and edges represent the relationships between these entities. Edges can be directed or undirected and can have properties that describe the relationship between the nodes. For example, a social network can be represented as a graph where each user is a node, and the edges represent the relationships between users, such as "friend" or "follow". The edges can also have properties, such as the date the relationship was formed or the strength of the relationship. The graph data model is often used in applications such as social networks, recommendation systems, and network analysis. It allows for efficient querying and analysis of complex relationships and patterns within the data.

Gephi is an open-source software application used for graph visualization and analysis. It provides a user-friendly interface for creating and manipulating graph data models [6]. To create a graph data model with Gephi, one can follow these steps:

- *Import Data*: You can import data in various formats, such as CSV, Excel, or a database. In the "Data Laboratory" tab of Gephi, you can add nodes and edges by importing data files.
- *Define Nodes and Edges*: In the "Data Laboratory" tab, you can define the attributes of nodes and edges. For example, you can add attributes such as name, age, and gender to nodes and attributes such as weight, type, and direction to edges.
- *Visualize the Graph*: Once you have defined the nodes and edges, you can visualize the graph by switching to the "Overview" tab in Gephi. In this tab, you can customize the appearance of nodes and edges, change the layout, and adjust the graph's overall appearance.
- *Analyze the Graph*: Gephi provides several analysis tools that can help you understand the structure of the graph. For example, you can run community detection algorithms to identify groups of nodes with similar characteristics or run centrality measures to identify nodes with the most influence in the graph.

Gephi provides a powerful set of tools for creating and analyzing graph data models, making it a popular choice for researchers and analysts working with network data.

## 10.1.4 Data Management Tools

Data management tools help organizations manage their data effectively by providing solutions for data integration, data quality, data governance, and data security [7,8]. Some of the popular data management tools from the perspective of big data are discussed in the following sections.

## 10.1.4.1 *Redis*

Remote Dictionary Server (Redis) is an open-source, in-memory data structure store that can be used as a database, cache, and message broker [9]. It is designed to provide high performance, scalability, and reliability for applications that require fast data access and low latency. Redis is a key-value store that supports a wide range of data structures, including strings, hashes, lists, sets, and sorted sets. It also supports advanced features such as transactions, pub/sub messaging, Lua scripting, and clustering.

Redis can be used as a cache to store frequently accessed data in memory, which can significantly improve application performance. It can also be used as a database to store and retrieve data, especially for applications that require high-speed data access. In addition to caching and database use cases, Redis can also be used as a message broker for building real-time applications. Redis provides a pub/sub messaging model that allows multiple clients to subscribe to specific channels and receive real-time updates when data changes.

Redis is widely used by many organizations, including GitHub, Twitter, Craigslist, and Stack Overflow. It is available in both open-source and commercial versions, and it can be deployed on-premises or in the cloud. Redis is also supported by a large and active community of developers who contribute to its development and provide support to users.

## 10.1.4.2 *Aerospike*

Aerospike is a high-performance, distributed NoSQL database that is designed to provide low-latency data access and high throughput for real-time applications. It is a key-value store that is optimized for large-scale, mission-critical applications that require high availability and predictable performance [10]. Aerospike is designed to operate in a distributed cluster of nodes, which provides high availability, fault tolerance, and scalability. It supports strong consistency and eventual consistency models, and it provides tunable consistency settings to allow applications to balance consistency and performance requirements.

Aerospike is a memory-first database, which means that it stores data primarily in memory to provide fast data access and low latency. It also provides options for persistent storage, which allows data to be stored on disk for durability and backup purposes. Aerospike supports a wide range of data structures, including strings, integers, lists, maps, and sets. It also supports advanced features such as secondary indexes, aggregations, and queries.

Aerospike is used by many organizations for real-time applications, such as ad tech, e-commerce, financial services, and online gaming. It is available in both open-source and commercial versions, and it can be deployed on-premises or in the cloud. Aerospike is also supported by a large and active community of developers who contribute to its development and provide support to users.

### 10.1.4.3 *AsterixDB*

AsterixDB is an open-source big data management system that is designed to support semi-structured data. It provides a platform for storing, managing, and querying large volumes of data that are complex and heterogeneous in nature [11,12]. AsterixDB is built on top of the Apache Hadoop ecosystem, and it is designed to work with a wide range of data formats, including JSON, XML, CSV, and other semi-structured formats. It provides a declarative language called Asterix Query Language (AQL) for querying and processing data, which supports a wide range of data manipulation operations, such as filtering, grouping, and aggregation.

AsterixDB is optimized for both batch and real-time data processing, and it provides a scalable and fault-tolerant architecture that can be deployed on-premises or in the cloud. It supports the distributed processing of data across multiple nodes, and it provides features such as data partitioning, replication, and load balancing. AsterixDB also provides support for advanced analytics, including text search, graph analysis, and machine learning. It provides integration with popular machine learning frameworks, such as TensorFlow and PyTorch, allowing users to train and deploy machine learning models on large datasets.

AsterixDB is used by many organizations in industries such as finance, healthcare, and telecommunications. It is available as an open-source software under the Apache License, and it is supported by a large and active community of developers who contribute to its development and provide support to users.

## 10.1.4.4 *Solr*

Solr is an open-source search platform that provides powerful indexing and search capabilities for websites, applications, and data stores [13]. It is built on top of the Apache Lucene search library, which provides a rich set of indexing and search features. Solr provides a RESTful API for managing and querying search indexes, which supports a wide range of search features such as full-text search, faceted search, and geospatial search. It supports a wide range of data formats, including JSON, XML, and CSV [14].

Solr is designed to be highly scalable and fault-tolerant, and it supports distributed indexing and search across multiple nodes. It also provides features such as replication, sharding, and load balancing, which allow it to handle large volumes of data and queries. Solr is often used as a search engine for websites and e-commerce platforms, as well as for search-based applications, such as customer service portals and recommendation engines. It is also used for data discovery and exploration in big data environments, where it can be used to search and analyze large volumes of data.

Solr is available as an open-source software under the Apache License, and it is supported by a large and active community of developers who contribute to its development and provide support to users.

## 10.1.4.5 *Vertica*

Vertica is a high-performance, column-oriented relational database management system (RDBMS) that is designed to provide fast and scalable data analytics for large-scale data warehouses and big data applications. It is built on a massively parallel processing (MPP) architecture, which allows it to process and analyze large volumes of data in parallel across multiple nodes [15,16]. Vertica is optimized for complex queries and analytics, and it supports a wide range of SQL functions, including window functions, analytical functions, and aggregate functions. It also provides advanced analytics features, such as machine learning algorithms and geospatial analysis.

Vertica is highly scalable and can be deployed on-premises or in the cloud. It supports data compression and partitioning, which allows it to efficiently store and manage large volumes of data. It also provides features such as load balancing, failover, and disaster recovery, which ensure high availability and reliability. Vertica is often used for data warehousing and big data analytics in industries such as finance, healthcare, and e-commerce. It is available in both open-source and commercial versions, and it is supported by a large and active community of developers who contribute to its development and provide support to users.

## 10.2  Big Data Integration and Processing Tools

*Big data integration* refers to the process of combining and transforming large volumes of data from various sources into a single, unified view [17]. This process is often necessary because big data is typically stored in multiple locations and in different formats, making it difficult to analyze and derive insights from.

The goal of big data integration is to provide a comprehensive view of an organization's data assets, allowing analysts and decision-makers to extract insights and make data-driven decisions. The process typically involves several stages, including:

- *Data Ingestion*: This involves collecting data from various sources, such as databases, APIs, and streaming platforms, and transferring it to a central location for processing and analysis.
- *Data Transformation*: This involves cleaning, enriching, and normalizing the data to ensure it is accurate, consistent, and compatible with other data sources.
- *Data Storage*: This involves storing the integrated data in a data warehouse or data lake, where it can be easily accessed and analyzed.
- *Data Processing*: This involves using big data processing tools, such as Hadoop, Spark, and Flink, to perform advanced analytics on the integrated data.
- *Data Visualization*: This involves presenting the integrated data in a way that is easy to understand and actionable using tools such as dashboards and reports.

Big data integration can be challenging due to the large volume, variety, and velocity of the data involved. However, with the right tools and techniques, organizations can successfully integrate their big data and derive valuable insights to support their business objectives.

There are several big data integration tools available on the market that can help organizations with their data integration needs. Some of the popular ones include:

- *Apache NiFi*: An open-source data integration tool that provides an easy-to-use web interface for designing, building, and managing data flows. It can handle large volumes of data and supports real-time data streaming [18].
- *Talend*: An open-source data integration tool that supports a wide range of data sources and destinations, including big data platforms such as Hadoop and Spark. It provides a visual interface for designing data integration workflows and supports both batch and real-time data processing [19].
- *Informatica*: A commercial data integration tool that provides a wide range of data integration capabilities, including data profiling, data cleansing, and data quality management [20]. It supports a wide range of data sources and destinations, including big data platforms such as Hadoop and Spark.
- *Apache Kafka*: An open-source data streaming platform that provides a distributed messaging system for real-time data processing. It can handle large volumes of data and provides high throughput and low latency [21].
- *Apache Spark*: An open-source big data processing engine that provides a wide range of data processing capabilities, including batch processing, stream processing, machine learning, and graph processing [22]. It can integrate with a wide range of data sources and support multiple programming languages.

Big data integration tools can help organizations with their data integration needs and enable them to process and analyze large volumes of data more effectively and efficiently.

## 10.2.1  Big Data Processing Using Splunk and Datameer

Splunk and Datameer are two popular big data processing tools used for analyzing and visualizing large datasets [23]. While they have some similarities, they have different features and use cases.

Splunk is a powerful tool for analyzing and visualizing machine-generated data, such as logs, events, and metrics. It provides a real-time search and analysis platform that makes it easy to explore and understand large volumes of data. Some key features of Splunk include:

- *Data Ingestion*: Splunk can ingest data from various sources, such as logs, sensors, and social media, in real-time or batch mode.
- *Search and Analysis*: Splunk provides a search language that allows you to query and analyze the data. It also provides a wide range of visualizations and dashboards for exploring the data.
- *Machine Learning*: Splunk provides machine learning capabilities for anomaly detection, predictive analytics, and other advanced use cases.
- *Integration*: Splunk can be integrated with other big data tools and platforms, such as Hadoop, Spark, and Kafka, to perform more complex analyses on the data.

Datameer is a big data analytics platform that provides a wide range of data integration, preparation, and analysis tools. It is designed to handle large and complex datasets and provide an easy-to-use interface for data analysts and business users. Some key features of Datameer include:

- *Data Integration*: Datameer can ingest data from various sources, such as Hadoop, databases, and cloud services, and transform them into structured data for analysis.
- *Data Preparation*: Datameer provides a wide range of data preparation tools, such as data profiling, cleansing, and enrichment, that allow you to clean and enrich the data before analysis.
- *Data Analysis*: Datameer provides a wide range of data analysis tools, such as statistical analysis, predictive modeling, and machine learning, that allow you to explore and understand the data.
- *Visualization*: Datameer provides a wide range of visualizations and dashboards for exploring and presenting the data.
- *Integration*: Datameer can be integrated with other big data tools and platforms, such as Hadoop, Spark, and Hive, to perform more complex analysis on the data.

Both Splunk and Datameer are powerful tools for processing and analyzing large datasets. While Splunk is focused on analyzing machine-generated data in real time, Datameer provides a wider range of data integration and preparation tools for business analysts and data scientists.

## 10.3 Big Data Machine Learning Tools

This section addresses two prominent tools: KNIME [24] and Spark MLlib [25]. Both of these are examples of open-source software. KNIME is a graphical user interface (GUI)-based tool, whereas Spark MLlib is a programmable platform for handling extremely big datasets.

### 10.3.1 KNIME

Konstanz Information Miner (KNIME) is an open-source data analytics, reporting, and integration platform developed by KNIME AG, a software company based in Konstanz, Germany [24]. KNIME allows users to visually create data workflows, which can include tasks such as data acquisition, transformation, analysis, and visualization.

Nodes are the names given to the fundamental building blocks in KNIME. KNIME provides a wide range of built-in nodes for data manipulation, analysis, and machine learning, as well as support for integrating external tools and libraries. Users can also create their own nodes using programming languages, such as Java or Python. KNIME workflows can be saved, reused, and deployed as web services or batch processes. In order to generate a workflow, the user must first select relevant nodes from the node repository and then put those nodes together into a workflow, as shown in Figure 10.1. After that, the workflow can be run on the KNIME workbench. Each workflow operation is implemented by a node in the network.

In this image, there are several nodes. One of these nodes is known as the file reader node, and its purpose is to extract data from a website link, a URL, or even a text file. The other node, known as the decision tree learner node, is responsible for creating a model of a decision tree.

Each node has access to other nodes' input and output ports, and they can communicate with one another, as shown in Figure 10.2.

The execution of a node involves reading data from the node's input port, processing that data, and then writing the results of that processing to the node's output port. There is a transmission of data between the nodes that are coupled together. Opening a node's setup dialogue is

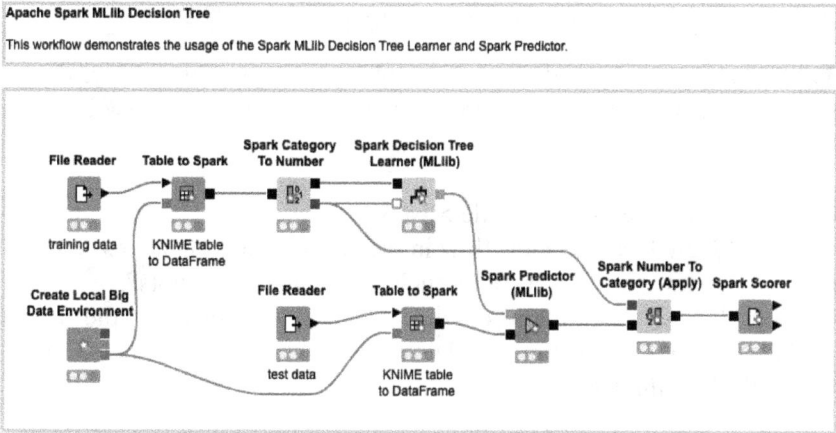

**Figure 10.1.** Spark MLlib Decision Tree Workflow [26]

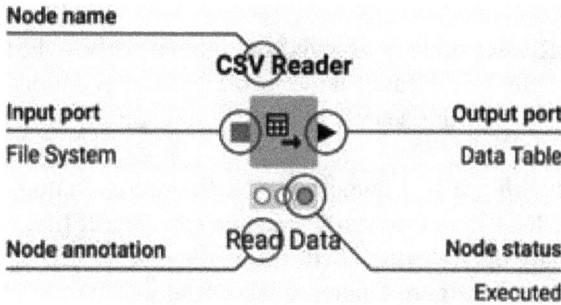

**Figure 10.2.** Node Structure and Status [27]

required before you may configure it. The parameters of the node can be modified in this section. The Node Repository stores all of the nodes that are compatible with a particular KNIME installation. Categories are used to arrange the nodes in the repository. KNIME offers a wide variety of nodes that can be used for a variety of data-related tasks, including accessing, manipulating, analyzing, visualizing, and reporting. KNIME offers a graphical interface for machine learning, which has a drag-and-drop interface that facilitates the creation and execution of machine learning workflows. However, the open-source version of KNIME can only handle limited-sized datasets. There are paid modules available for KNIME that can manage extensive datasets and provide additional functionality.

### 10.3.1.1  *Exploring Data with KNIME Plots*

Before one can import a dataset and explore it, we first need to creating a KNIME workflow. Following are the key steps to create a new KNIME workflow:

- In the main menu, select File > New.
- Select New KNIME Workflow in the window and click Next >.
- Give a name to the workflow, such as "Exploratory Plots".
- The default destination of the KNIME workflow is the Local Workspace. Click Complete. The new workflow should appear under LOCAL in the KNIME Explorer view.

After creating the workflow, the next task is to import the dataset. For further analysis, we consider the daily_weather.csv [28] dataset. Following steps are followed to import the dataset:

- The File Reader node is dragged onto the Workflow Editor. You can search for the File Reader node in the Node Repository by entering "File Reader" in the search box or by navigating to the IO > Read category.
- Double-clicking it will launch the Configuration Dialog.
- Select the location of the daily_weather.csv dataset file, which should have already been downloaded.
- To close the Configure Dialog, click OK.
- Right-click a node and select Execute.

Now, we have created a workflow, and the dataset is also imported. We then see different exploratory tools.

*Histogram*: The examination of the distribution of a continuous quantity is what a histogram is designed to do. It does it by grouping the data into containers called *bins* and then plotting the occurrence frequency within each bin's respective range. We visualize the data represented by the air_temp_9am column.

Following are the steps to create a histogram, as shown in Figure 10.3:

- Inside the Node Repository, navigate to the "Views" section to locate the Histogram node, and then drag it into the Workflow Editor.

**File Reader**                    **Histogram**

daily_weather.csv                  Histogram of
                                   air_temp_9am

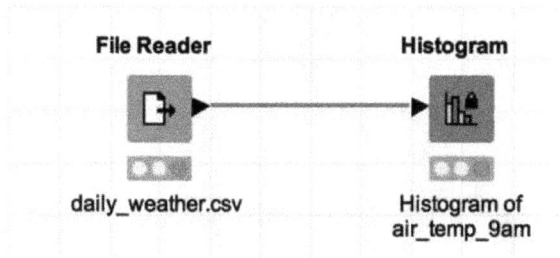

**Figure 10.3.**   Creation of Histogram in KNIME [28]

- Create a connection between the Histogram and File Reader nodes.
- Double-clicking on the Histogram node will bring up the Configure dialogue for you to work with. Make sure that the air_temp_9am column is selected for both the Binning Column and the Aggregation Column. The Number of Bins can be set to anything that makes sense, taking the data into account. The benchmark is 10 points.
- Click the OK button to commit these changes and end the configuration dialogue.
- To begin the workflow, you must first double-click the green arrow that is located at the very top of the screen.
- To see the air_temp_9am histogram, right-click on the Histogram node, then go to the context menu and select "View: Histogram View". The labels along the *x*-axis show the temperature range that each bin contains, while the labels along the *y*-axis show how often samples are taken in each bin.

The distribution of the temperature at 9 a.m. is normal (bell-shaped), according to the histogram shown in Figure 10.4. The temperature ranges from 34 to 99 degrees Fahrenheit, with frequent readings between 60° and 73°. No anomalies warrant concern. The final bin on the right indicates the number of samples with missing values.

*Scatter Plot*: A scatter plot is a graphical representation of data that shows the relationship between two variables by plotting them as points on a graph (Figure 10.5).

The scatter plot shown in Figure 10.6 shows that there is a negative correlation between temperature and relative humidity.

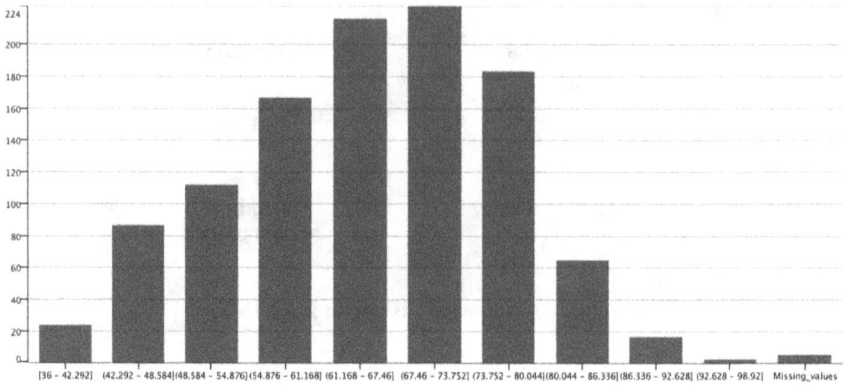

**Figure 10.4.**   Histogram for Air Temperature [28]

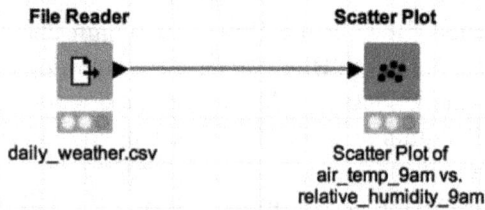

**Figure 10.5.**   Creation of Scatter Plot in KNIME [28]

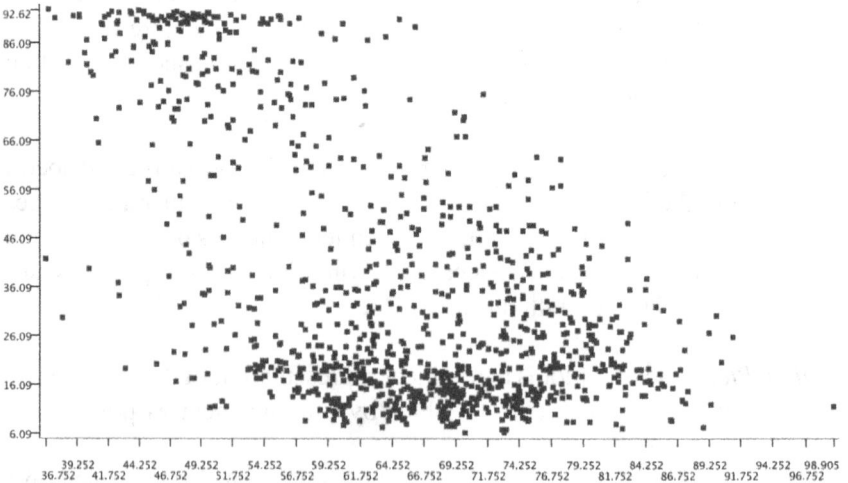

**Figure 10.6.**   Scatter Plot for Relative Humidity

This means that an increase in temperature corresponds to a decrease in relative humidity. Since warm air can hold more water vapor than cool air, relative humidity falls when the temperature rises. This relationship is captured by the scatter plot.

*Box Plot*: One can make comparisons between different distributions by using a box plot. The data for a numeric variable are first organized into groups, and then, a box plot is produced to show the distribution of each category's values. After that, the box plots are shown on a single graph so that the various categories may be compared side by side.

It will create a box plot to examine how the distribution of air pressure at 9 a.m. differs between days with low humidity, normal humidity, and high humidity. First, a categorical variable named "low humidity day" will be created to indicate whether a day has low humidity or not. This is possible with the Numeric Binner node, as shown in Figure 10.7. This variable's condition is "if relative humidity 9 a.m. is less than 25% then low humidity day = 1; otherwise low humidity day = 0".

The box plot depicted in Figure 10.8 indicates that, on average, pressure is greater on days with low humidity. Low-pressure weather systems are associated with stormy and rainy conditions (with high humidity), whereas high-pressure systems bring clear skies and low humidity.

## 10.3.1.2 *Handling Missing Values in KNIME*

While creating the histogram for air temperature, we observed missing values in the last bin. One way to handle the missing values is to simply

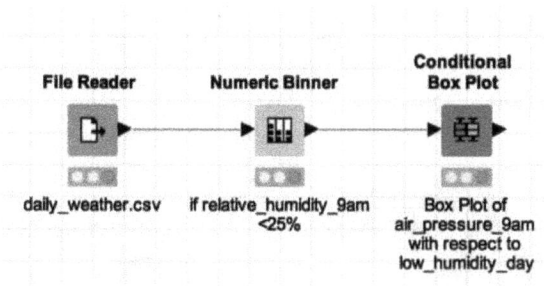

**Figure 10.7.** Creation of Box Plot in KNIME [28]

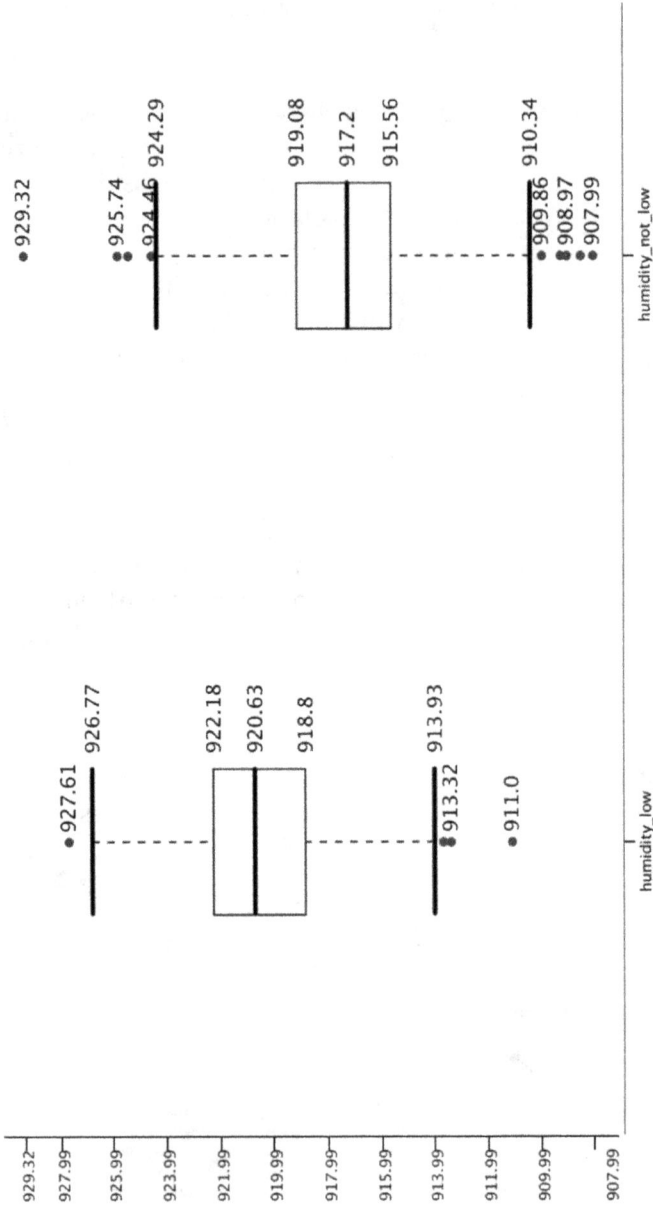

**Figure 10.8.**   Conditional Box Plot for Relative Humidity [28]

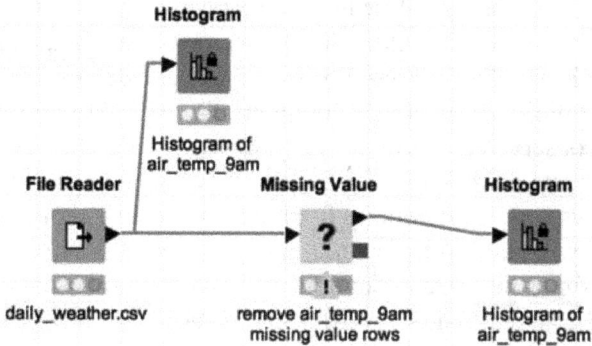

**Figure 10.9.** Removing Missing Values in KNIME [28]

remove the row that contains them. Following are the steps in KNIME to remove the missing values, as shown in Figure 10.9:

- Search for "Missing Value" in the Node Repository and drag the Missing Value node onto the Workflow Editor. Connect the Node with Missing Value to the Node with File Reader.
- In the Missing Value node's Configure dialogue, navigate to the Column Settings tab. Select the air_temp_9am column, click the Add button, and then select Remove Row* from the drop-down menu. This means that samples with a missing air_temp_9am value will be discarded.
- Copy the Histogram node by right-clicking and pasting it to the right of the Missing Value node. Connect the Missing Value node's black triangle output to the input of the second Histogram node.

*Input Missing Values with Mean*: Another method for handling missing values is to replace them with the column's mean or median. This can be accomplished using the existing Missing Value node pipeline.

### 10.3.1.3 *Classification Using Decision Tree in KNIME*

After exploring the data and examining how to deal with missing values, the next step is to construct a classification model to predict days with low humidity. Remember that low humidity is one of the weather conditions that increases the risk of wildfires, so it would be advantageous to be able to forecast days with low humidity. We construct a decision model to classify low-humidity days vs. non-low-humidity days based on the weather conditions observed at 9 a.m.

Let's create a workflow for developing a decision tree model to distinguish days with low humidity from those with normal or high humidity. The model will be used to forecast days with low humidity.

Apart from the three nodes that we have already discussed, we have now added a Statistics node to check certain aspects of the processed data, as shown in Figure 10.10. Following are the steps followed in performing statistical analysis:

- Connect a Statistics node to the File Reader node's output. Change the Max number of possible values per column (in the output table) in the Configure Dialog of the Statistics node to 1,500 and add all >> columns to the Include side. This node should be renamed "Statistics BEFORE filtering".
- Connect a Statistics node to the Numeric Binder node's output. Change the Max number of possible values per column (in the output table) in the Configure dialogue of this Statistics node to 1,500 and add all >> columns to the Include side. Rename this node "Statistic AFTER Filtering".

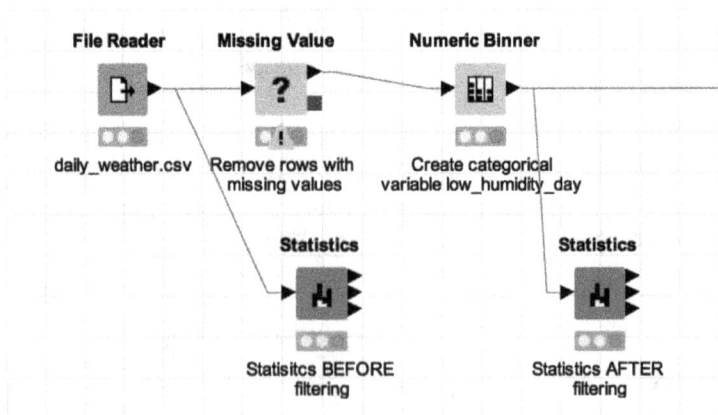

**Figure 10.10.**   Prepare Data for Classification [28]

- When both Statistics nodes are executed and viewed, the resulting histograms should have identical characteristics. This is to ensure that our treatment of missing values does not skew our data.

Now, we can build a Decision Tree workflow, as shown in Figure 10.11, as follows:

- Join a Column Filter to a Numeric Binner. In the Column Filter node's Configure Dialog, only the relative_humidity_9am and relative_humidity_3pm columns should be excluded.
- Connect a node from the Color Manager to the Column Filter node. This will color-code the categorical variable, low_humidity_day, making it easier to visualize in subsequent workflow steps. Check in the Configure Dialog of the Color Manager node that humidity_low is colored red and humidity_not_low is colored blue for low-humidity days.
- Connect a node for Partitioning to the node for Color Manager. In order to split the data into a training piece and a testing section, you will need the Partitioning node. The training data are utilized in the construction of the decision tree, while the test data are utilized to evaluate the classifier's performance on data that have not previously been seen. Make your selections in the Configure Dialog of the Partitioning node as follows: Relative [%] 80, Draw Randomly, and Use a random seed. Because of this, 80% of the data will be allocated to the first output, which is the training data, and 20% will be

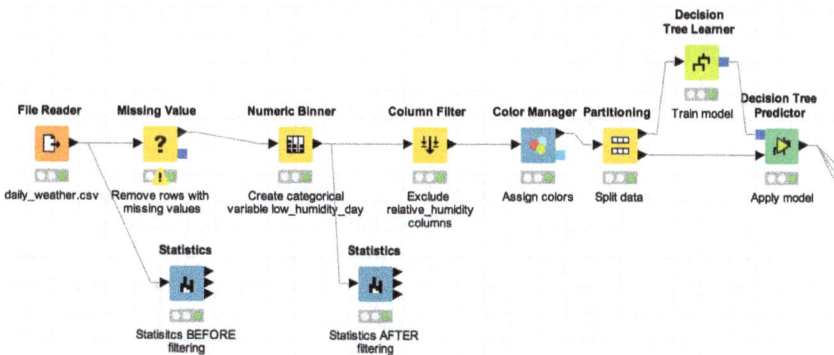

**Figure 10.11.** Decision Tree Workflow in KNIME [28]

allocated to the second output (the test data). The random seed is placed in this location to ensure that everyone receives identical training and test datasets for the purpose of training and testing the decision tree model.

- Make a connection between the first output of the Partitioning node and the Learner node of the Decision Tree. This node is responsible for generating the classification decision tree by making use of the training data. In the Configure dialogue, reduce the minimum number of records required for each node to 20. This is a stopping criterion for the algorithm for tree induction. It specifies that a node with this many samples cannot be split anymore. This parameter's default value is 2, which is very small and may lead to overfitting.
- Join the outputs of the Partitioning node and the Decision Tree Learner node with a Decision Tree Predictor. The model will be applied to the test data by this node.
- Carry out the workflow. Choose "View: Decision Tree View" from the context menu that appears when you right-click the Decision Tree Predictor node to see the created decision tree.

### 10.3.1.4 *Evaluation of Decision Tree in KNIME*

In this section, we execute the following KNIME operations:

- Establish and analyze a confusion matrix for a decision tree.
- Assess the accuracy of a decision tree model.

*Generation of Confusion Matrix*: A confusion matrix displays the classification errors and correct classifications made by a classifier. It can be generated using a Scorer node as follows:

- Launch the Decision Tree Workflow created in the Classification Hands-On reading.
- Connect a Scorer node to the existing Predictor node of the Decision Tree.
- By default, the Scorer Configure Dialog should appear as shown. Click Accept.
- Perform and view the Scorer operation. It displays the confusion matrix and the precision of the prediction.

## 10.3.2 Spark MLlib

The machine learning library MLlib is built on top of Spark, making it very scalable. Common machine learning techniques and tools are distributed and implemented by it [25]. To execute machine learning tasks with Spark MLlib, one needs to write code, as the library does not provide a graphical user interface. The MLlib library, like Spark itself, provides an API in the languages Java, Python, Scala, and R. This allows programs developed in these languages to utilize Spark's core functionality. Spark MLlib uses a decentralized system to function. It offers machine learning algorithms and methods that can be used with distributed computing. Consequently, MLlib is used to process and analyze large datasets.

# 10.4 Big Data Graph Analytics Tools

In this section, we discuss two dominant systems developed for large-scale graph processing. The first is called Giraph, which is developed by Apache and implements a bulk synchronous parallel (BSP) model on Hadoop [29]. The second system, known as Graphx, is built on the Spark platform, which prioritizes interactive in-memory computations [30]. BSP is a popular graph processing model, but the actual implementation of BSP in infrastructure requires additional programmability. Before we can discuss how these tools are used for analytics purposes, we first have to consider the graph IO, i.e., how graphs can come into a system represented inside the system and, when completed, be written out into files. A graph can be constructed in numerous ways. Importing graphs from a CSV file into the database is possible with Neo4J.

The two most common input formats in Giraph are Adjacency List and Edge List, as shown in Figure 10.12. Each line of an Adjacency List contains the node ID, a single-digit node value, and a list of destination–weight pairs. A has a value of 10 and two neighbors with edge weights of 2 and 5, respectively, including B and F. In the edge list, graphs are represented in terms of triplets, containing the source and target nodes and the edge weight.

## 10.4.1 Giraph

Giraph is an open-source Apache project that provides a distributed graph processing framework for processing large-scale graph data. It is built on

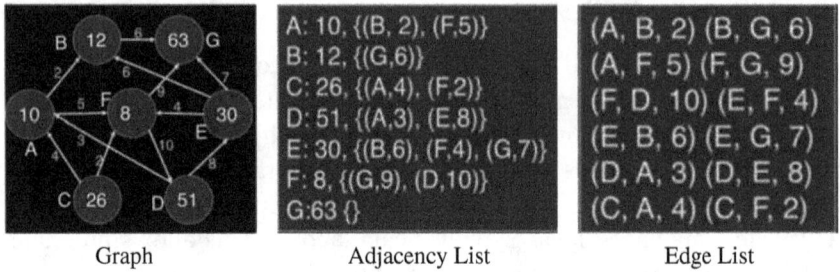

| Graph | Adjacency List | Edge List |
|---|---|---|
|  | A: 10, {(B, 2), (F,5)}<br>B: 12, {(G,6)}<br>C: 26, {(A,4), (F,2)}<br>D: 51, {(A,3), (E,8)}<br>E: 30, {(B,6), (F,4), (G,7)}<br>F: 8, {(G,9), (D,10)}<br>G:63 {} | (A, B, 2) (B, G, 6)<br>(A, F, 5) (F, G, 9)<br>(F, D, 10) (E, F, 4)<br>(E, B, 6) (E, G, 7)<br>(D, A, 3) (D, E, 8)<br>(C, A, 4) (C, F, 2) |

**Figure 10.12.** Graph Representation by Adjacency List and Edge List

top of Apache Hadoop, which provides a distributed storage system (HDFS) and a batch processing framework (MapReduce). Giraph was initially developed by Yahoo! in 2008 as an internal project and was later open-sourced in 2011. It is written in Java and provides a simple programming model for processing large-scale graphs in a distributed manner.

Some of the key features of Giraph include:

- support for various graph algorithms, such as PageRank, Connected Components, and Shortest Paths;
- support for graph mutation and partitioning;
- fault tolerance and automatic checkpointing;
- integration with Apache Hadoop and HBase;
- compatibility with various programming languages, such as Java, Python, and Scala;
- support for graph visualization.

Giraph is widely used in various industries, such as social media, finance, and e-commerce, for a wide range of use cases, including recommendation systems, fraud detection, and network analysis.

## 10.4.2 GraphX

GraphX is a distributed graph processing framework developed by Apache that provides an API for processing large-scale graph data. It is built on top of Apache Spark, which is a distributed processing engine that provides in-memory data processing capabilities. GraphX was first introduced in 2014 and is written in Scala. It provides a set of high-level

APIs for creating, transforming, and querying graphs in a distributed manner.

Some of the key features of GraphX include:

- support for various graph algorithms, such as PageRank, Connected Components, and Triangle Counting;
- support for graph mutation and partitioning;
- integration with Apache Spark's resilient distributed datasets;
- compatibility with various programming languages, such as Scala, Java, and Python;
- support for graph visualization.

GraphX is widely used in various industries, such as social media, finance, and e-commerce, for a wide range of use cases, including recommendation systems, fraud detection, and network analysis.

### 10.4.3 Neo4j

Neo4j is a graph database management system that allows users to store and retrieve data in the form of nodes, edges, and properties. Unlike traditional relational databases, where data are stored in tables, graph databases store data in a graph format, which allows for more flexible and efficient querying and analysis of interconnected data. Neo4j was developed by Neo4j, Inc. and released in 2010 as an open-source software under the GPL license [31]. It is written in Java and is compatible with various programming languages, such as Python, Ruby, and JavaScript.

Some of the key features of Neo4j include:

- high performance and scalability,
- ACID compliance,
- support for Cypher query language,
- support for clustering and replication,
- full-text search capabilities,
- integration with various programming languages and frameworks.

Graph analytics with Neo4j involves using Neo4j's graph database to perform analytical tasks on graph data. Neo4j provides several built-in algorithms and tools for graph analytics, as well as support for creating custom algorithms.

Some of the built-in graph analytics algorithms provided by Neo4j include:

- *PageRank*: A popular algorithm used for ranking web pages based on their importance.
- *Shortest Path*: An algorithm that finds the shortest path between two nodes in a graph.
- *Community Detection*: An algorithm that detects communities or clusters of nodes in a graph.
- *Betweenness Centrality*: An algorithm that measures the importance of nodes in a graph based on the number of shortest paths that pass through them.

Neo4j also provides a Graph Data Science library that includes additional algorithms and tools for graph analytics. This library includes algorithms such as Louvain Modularity, Label Propagation, and Node Similarity, as well as tools for graph embedding and graph visualization.

Graph analytics with Neo4j can be used in various industries, such as finance, healthcare, and social media, for a wide range of use cases, including fraud detection, recommendation systems, and knowledge graph management. It allows users to extract valuable insights from interconnected data and make data-driven decisions based on the results of their analysis.

## 10.5 Chapter Summary

The enabling tools and technologies are crucial for big data analytics, as they provide the necessary infrastructure and resources for managing and processing large volumes of data. With the rapid growth of data in today's world, it is essential to have efficient and scalable tools and technologies for analyzing, storing, and managing data.

The emergence of cloud computing, distributed systems, and NoSQL databases has revolutionized the way big data is handled and analyzed. These technologies provide scalable and cost-effective solutions for storing and processing data, enabling organizations to analyze large datasets in real time. Moreover, data visualization tools and machine learning algorithms are also important enabling technologies for big data analytics. Data visualization tools help transform complex data into visually appealing charts, graphs, and dashboards, allowing users to gain insights quickly and easily. Machine learning algorithms enable organizations to build predictive models and perform advanced analytics on their data, providing valuable insights and recommendations for decision-making.

In conclusion, enabling tools and technologies are essential for big data analytics, as they provide the foundation for managing, processing, and analyzing large volumes of data. As the volume and complexity of data continue to grow, it is important for organizations to stay abreast of the latest tools and technologies to remain competitive and gain valuable insights from their data.

# References

1. Vijayaraj, J., Saravanan, R., Paul, P. V., & Raju, R. (2016). A comprehensive survey on big data analytics tools. In *2016 Online International Conference on Green Engineering and Technologies (IC-GET)*, 19 November 2016, Coimbatore, India, (pp. 1–6). IEEE.
2. Pol, U. R. (2014). Big data and hadoop technology solutions with cloudera manager. *International Journal*, 4(11), 1028–1034.
3. Slavakis, K., Giannakis, G. B., & Mateos, G. (2014). Modeling and optimization for big data analytics: (statistical) learning tools for our era of data deluge. *IEEE Signal Processing Magazine*, 31(5), 18–31.
4. "Apache Lucene Documentation". https://lucene.apache.org/core/documentation.html.
5. "Lucene in Action book by Erik Hatcher and Otis Gospodnetić". https://www.manning.com/books/lucene-in-action.
6. Bastian, M., Heymann, S., & Jacomy, M. (2009). Gephi: An open source software for exploring and manipulating networks. In *Proceedings of the international AAAI Conference on Web and Social Media*, 5–8 June 2023, Limassol, Cyprus, (Vol. 3, No. 1, pp. 361–362).
7. Chaudhary, P., Gupta, B. B., & Singh, A. K. (2022). XSS Armor: Constructing XSS defensive framework for preserving big data privacy in internet-of-things (IoT) networks. *Journal of Circuits, Systems and Computers*, 31(13), 2250222.
8. Stergiou, C., Psannis, K. E., Xifilidis, T., Plageras, A. P., & Gupta, B. B. (2018). Security and privacy of big data for social networking services in cloud. In *IEEE INFOCOM 2018-IEEE Conference on Computer Communications Workshops (INFOCOM WKSHPS)*, 15–19 April 2018, Honolulu, HI, USA, (pp. 438–443). IEEE.
9. Carlson, J. (2013). *Redis in Action*. Simon and Schuster.

10. Ahmad, K. & Kamal, A. (2017). Hands-On Aerospike. In *NoSQL: Database for Storage and Retrieval of Data in Cloud* (pp. 311–322). Chapman and Hall/CRC.

11. Alsubaiee, S., Altowim, Y., Altwaijry, H., Behm, A., Borkar, V., Bu, Y., Carey, M., Cetindil, I., Cheelangi, M., Faraaz, K., Gabrielova, E., Grover, R., Heilbron, Z., Kim, Y. S., Li, C., Li, G., Ok, J. M., Onose, N., Pirzadeh, P., Tsotras, V., Vernica, R., Wen, J., Westmann, T. (2014). AsterixDB: A scalable, open source BDMS. *arXiv preprint arXiv:1407.0454.*

12. Alsubaiee, S., Behm, A., Borkar, V., Heilbron, Z., Kim, Y. S., Carey, M. J., Dreseler, M. & Li, C. (2014). Storage management in AsterixDB. In *Proceedings of the VLDB Endowment*, 7(10), 841–852.

13. Grainger, T. & Potter, T. (2014). *Solr in Action.* Manning Publications Co., Shelter Island, New York.

14. Smiley, D., Pugh, E., Parisa, K., & Mitchell, M. (2015). *Apache Solr Enterprise Search Server.* Packt Publishing Ltd., Birmingham, UK.

15. Lamb, A., Fuller, M., Varadarajan, R., Tran, N., Vandier, B., Doshi, L., & Bear, C. (2012). The vertica analytic database: C-store 7 years later. *arXiv preprint arXiv:1208.4173.*

16. Bear, C., Lamb, A., & Tran, N. (2012). The vertica database: Sql rdbms for managing big data. In *Proceedings of the 2012 Workshop on Management of Big Data Systems*, 21 September 2012, San Jose, California, USA, (pp. 37–38).

17. Patel, J. (2019). Bridging data silos using big data integration. *International Journal of Database Management Systems*, 11(3), 1–6.

18. "Apache NiFi Documentation". https://nifi.apache.org/docs.html.

19. Sreemathy, J., Nisha, S., & Gokula Priya, R. M. (2020). Data integration in ETL using TALEND. In *2020 6th International Conference on Advanced Computing and Communication Systems (ICACCS)*, 6–7 March 2020, Coimbatore, India, (pp. 1444–1448). IEEE.

20. "Informatica Cloud Data Integration User Guide". https://docs.informatica.com/data-integration/cloud-data-integration.html.

21. Le Noac'H, P., Costan, A., & Bougé, L. (2017). A performance evaluation of Apache Kafka in support of big data streaming applications. In *2017 IEEE International Conference on Big Data (Big Data)*, 11–14 December 2017, Boston, MA, USA, (pp. 4803–4806). IEEE.

22. Salloum, S., Dautov, R., Chen, X., Peng, P. X., & Huang, J. Z. (2016). Big data analytics on Apache Spark. *International Journal of Data Science and Analytics*, 1, 145–164.

23. Chandio, A. A., Tziritas, N., & Xu, C. Z. (2015). Big-data processing techniques and their challenges in transport domain. *ZTE Communications*, 1(10), 1–21.

24. Berthold, M. R., Cebron, N., Dill, F., Gabriel, T. R., Kötter, T., Meinl, T., Ohl, P., Thiel, K. & Wiswedel, B. (2009). KNIME-the Konstanz information

miner: Version 2.0 and beyond. *AcM SIGKDD Explorations Newsletter*, 11(1), 26–31.

25. Meng, X., Bradley, J., Yavuz, B., Sparks, E., Venkataraman, S., Liu, D., Freeman, J., Tsai, D., Amde, M., Owen, S., Xin, D., Franklin, M.J., Zadeh, R., Zaharia, M. & Talwalkar, A. (2016). Mllib: Machine learning in apache spark. *The Journal of Machine Learning Research*, 17(1), 1235–1241.

26. "Spark MLlib Decision Tree". https://hub.knime.com/knime/spaces/ Examples/latest/10_Big_Data/02_Spark_Executor/01_Spark_MLlib_ Decision_Tree~YOap5g_GXCzsyyuB.

27. "KNIME Getting Started Guide". https://www.knime.com/getting-started-guide.

28. https://github.com/words-sdsc/coursera/blob/master/big-data-4/daily_ weather.csv.

29. Sakr, S., Orakzai, F. M., Abdelaziz, I., & Khayyat, Z. (2016). *Large-scale Graph Processing Using Apache Giraph*. Cham: Springer International Publishing.

30. Gonzalez, J. E., Xin, R. S., Dave, A., Crankshaw, D., Franklin, M. J., & Stoica, I. (2014). Graphx: Graph processing in a distributed dataflow framework. In *11th {USENIX} Symposium on Operating Systems Design and Implementation ({OSDI} 14)*, 6–8 October 2014, Broomfield, CO, (pp. 599–613).

31. Miller, J. J. (2013). Graph database applications and concepts with Neo4j. In *Proceedings of the Southern Association for Information Systems Conference* (Vol. 2324, No. 36), March 2013, Atlanta, GA, USA.

# Index

9 789811 257117